Underground Engineering

Underground Engineering

Planning, Design, Construction and Operation of the Underground Space

BAI YUN
College of Civil Engineering, Tongji University,
Shanghai, China

Academic Press is an imprint of Elsevier
125 London Wall, London EC2Y 5AS, United Kingdom
525 B Street, Suite 1650, San Diego, CA 92101, United States
50 Hampshire Street, 5th Floor, Cambridge, MA 02139, United States
The Boulevard, Langford Lane, Kidlington, Oxford OX5 1GB, United Kingdom

Notices

Knowledge and best practice in this field are constantly changing. As new research and experience broaden our understanding, changes in research methods, professional practices, or medical treatment may become necessary.

Practitioners and researchers must always rely on their own experience and knowledge in evaluating and using any information, methods, compounds, or experiments described herein. In using such information or methods they should be mindful of their own safety and the safety of others, including parties for whom they have a professional responsibility.

To the fullest extent of the law, neither the Publisher nor the authors, contributors, or editors, assume any liability for any injury and/or damage to persons or property as a matter of products liability, negligence or otherwise, or from any use or operation of any methods, products, instructions, or ideas contained in the material herein.

British Library Cataloguing-in-Publication Data
A catalogue record for this book is available from the British Library

Library of Congress Cataloging-in-Publication Data
A catalog record for this book is available from the Library of Congress

ISBN: 978-0-12-812702-5

For Information on all Academic Press publications
visit our website at https://www.elsevier.com/books-and-journals

Publisher: Matthew Deans
Acquisition Editor: Ken McCombs
Editorial Project Manager: Ali Afzal-Khan
Production Project Manager: Bharatwaj Varatharajan
Cover Designer: Matthew Limbert

Typeset by MPS Limited, Chennai, India

CONTENTS

PREFACE

Sustainable development involves meeting the needs of the present without compromising the ability of future generations to meet their own needs. To achieve the goal, human beings must further use underground spaces in an effort deal with rapid increase of population, expansion of urbanization, in addition to the effects of climate changes, to ensure resilience against natural disasters, preserve the environment, etc.

Underground engineering is an old subject. In fact, it originated when people lived in caves and scratched into the rock or stiff clay and dug the first underground structures. However, underground engineering is also a new technology, and its theories and methods are still in development. Today, underground construction involves different costs, ground conditions, cultural aspects, religious beliefs, as well and local and national political influences. From this point of view, underground engineering can also be considered as a discipline of art. Even in the 21st century, almost every underground project is a journey into the unknown as only about 0.1% of the ground is known before construction.

The 21st century is the century of underground engineering, and many cities in the world are excavating for subways, underground roads, utilities, water projects, sewage treatments, underground storage, underground plants, and other kinds of different underground facilities. For example, all in all more than 100 km of metro tunnel was being driven in Shanghai in 2015 alone. However, until now there have been few universities in the world teaching underground engineering. Today, Tongji University provides underground engineering courses for undergraduate students in the school of civil engineering, and also provides English language teaching for underground engineering, a need for which this text will fill. Although this book is a useful textbook undergraduate students, it is also a technical reference for young engineers engaged in underground engineering around the world.

Along with the development of underground engineering, the Muir Wood "spirit" can be understood as follows:

"Innovation in tunneling is key to economy and safety." Above all, successful tunneling depends on management of the uncertainty of the ground and how it can affect a specific project. The success of the tunneling scheme thus depends greatly on the competence of the engineer, including the ability to understand the owner's interests as well as the limitations and advantages of existing construction techniques. Engineering economy and efficiency free the contractor from needing to determine risks and understanding "reference conditions" that determine physical features and thus potential contractor liabilities. The secret of success in tunneling is recognizing the ubiquity of uncertainty involved in underground spaces. This uncertainty requires a management strategy specific to the project to minimize risk.

ACKNOWLEDGMENTS

The manuscript has been edited by Ms. Allisa Zhao, Mrs. Nicola MinnaFung, Miss. Jasmine, Miss. Diana Margarita Diaz, and Mr. Zhou Chen. Their comments were especially helpful. The author also appreciates the following students for their editing work:

Chapter 1: Mr. Xu Xiaofei and Mr. Xiao Li;

Chapters 2 and 3: Miss Zhao Bingyu;

Chapter 4: Miss PengJiamei, Mr. Lu Honghao, and Mr. Zhang Xuehui;

Chapter 5: Miss Li Yanxiang and Miss Wu Xiaoxiao;

Chapter 6: Mr. Lu Honghao;

General layout: Mr. Lu Honghao and Miss PengJiamei.

BRIEF INTRODUCTION

This book offers an overview of the field of underground engineering. After presenting the history of subsurface development and highlighting the goals for building underground structures, design and planning processes are discussed in detail. Numerous tunnel construction techniques and project management models are also covered. Lastly, operation systems disaster risks and protection measures are discussed to ensure project owners and managers are prepared for events that may jeopardize a tunnel project.

Dedicated to those new to underground engineering, this text aims to equip readers with a solid understanding of the field. In addition to giving an overview of underground engineering this resource all provides useful examples and case studies to facilitate understanding of practical aspects of subsurface structures and their design.

Current and innovative techniques and future trends are also discussed throughout to provide readers with the current state of the art. Each chapter concludes with recommendations on existing literature for readers that want to deepen their knowledge.

CHAPTER 1

History of Subsurface Development

Contents

Subsurface space is used in a variety of ways in the form of tunnels, mines, shelters, and burial chambers. For four millennia, human beings have dug tunnels and structures, some of which are still in use today and have thus stood the test of time.

The design and construction of subsurface structures have become increasingly more manageable and safer. But the complexity of the techniques used in underground engineering remains a challenge. However,

Underground Engineering
DOI: https://doi.org/10.1016/B978-0-12-812702-5.00001-3

construction methods, skills, and knowledge have evolved over the years, showing the importance of empirical learning in underground engineering. Indeed, developed technology can be refined thanks to long-term experience.

In this first chapter, a brief overview of subsurface structures is given, along with the different uses of subsurface space throughout history in varying geologies, cultures, and climates. This enumeration is not exhaustive and primarily focuses on Chinese examples.

1.1 CAVES AND GROTTOS

Over thousands of years, humans have been attached to the underground for many reasons, among others, for basic survival, artistic expression, and religious ceremonies. In the Stone Age, humans lived in caves (as confirmed by discovered cave paintings). More than 12,000 years ago, Stone Age men built, excavated, and extended tunnel networks, of which some parts still exist (Daily Mail Reporter, 2011). In the Chauvet-Pont-D'Are cave in southern France, the evocative paintings and engravings of animal and hunting scenes have been carbon-dated at more than 30,000 years old. Tunnels offer protection from predators.

Since the Stone Age, caves and grottos have been used for different purposes, some natural (Fig. 1.1) and others manmade. Some examples of caves and grottos are discussed here.

Figure 1.1 Nature cave in Shanxi Province, China.

1.1.1 Caves as Burial Sites

In the search for the beginning of underground construction, we begin in Ancient Egypt where proof of such structures still exists and has been well preserved and documented.

Many underground spaces were excavated as integral parts of pyramids or sacred burial sites. The most impressive pyramids are those of Giza, the biggest of the three being the Great Pyramid of Khufu. Fig. 1.2 shows the internal organization of the pyramid where different passages lead to the three known chambers: the lowest, the Queen's, and the King's chambers.

Another important example of an underground burial site is in Malta, namely the Hal Saflieni Hypogeum that dates from 2400 BC. It is a large underground temple consisting of a complex series of chambers and tombs (Fig. 1.3). The Hal Saflieni Hypogeum situated beyond the

Figure 1.2 Great Pyramid of Khufu in Giza, Egypt. *Adapted from Science. How Stuff Works. Freudenrich, 2007, http://science.howstuffworks.com/engineering/structural/pyramid2.htm.*

Figure 1.3 The Hal Saflieni Hypogeum, an underground cemetery (Gorski, 2006).

Figure 1.4 Outside the Mogao grottos (Cascone, 2014).

southern tip of Italy is comprised of three levels and reaches a depth of 12 m (Ring, Salkin, & La Boda, 1995).

1.1.2 Caves as Temples and Monasteries

Situated in Dunhuang in Gansu province, the Mogao Caves (also called Thousand Buddhas grottos) are one of the most valuable cultural heritage sites in China, containing exquisite paintings and sculptures (Fig. 1.4). With over 1600 years of history and more than 492 well-preserved Buddhist caves, the site was listed as a World Heritage site in 1987 (Unesco, n.d.). Since Dunhuang is located along the Silk Road, the caves were of great importance in artistic exchanges between China, India, and Central Asia.

1.1.3 Caves as Dwellings

In China since prehistoric times the Loess caves (the Loess Plateau [also called the Huangtu Plateau] is a 640,000 km^2-plateau covering much of Shanxi and Shaanxi as well as the northern Henan and eastern Gansu) and caves have traditionally been used as dwellings (Fig. 1.5). At the time caves were a dominant form of rural housing with an estimated 40 million people living in caves at the end of the last century. Cave dwellings are a successful ecological adaptation. They provide shelter, hold heat well, are affordable, and require no maintenance (Yoon, 1990).

1.2 ANCIENT MINES

Since the beginning of civilization humans have used materials found underground to build tools and weapons. For instance, flint was extracted

Figure 1.5 Loess caves in Zhangshanying, close to Beijing (taQpets, 2013).

Figure 1.6 Drawing of the Grime's graves (AncientCraft, n.d.).

through mining in chalk-rich areas and used to make tools. Used in the Neolithic Age, most flint mines date back to 4000 BC to 3000 BC (Crystalinks, n.d.). Fig. 1.6 shows the scheme of the Grime's caves in England, a 5000-year-old large flint-mining complex.

The oldest mining caves on archaeological record are located in Swaziland. Carbon dating has estimated the Lion cave to be more than 45,000 years old. At the time, people were interested in ore useful for traditional rituals (Swaziland National Trust Commission, n.d.). According to estimations, these mines were used until at least 23,000 BC.

Mining dates back thousands of years and has been employed by many of the great civilizations such as the Ancient Egyptians and Romans. The latter established large-scale hydraulic mining methods where numerous aqueducts were used to transport large volumes of water. This water had several purposes, such as removing rock debris, washing comminution ore, and powering simple machinery. For dewatering deep mines reverse overshot waterwheels were used at Rio Tinto (Crystalinks, n.d.).

Other examples of ancient mines can be found in South America, where they were used for the early extraction of precious minerals such as gold and silver, metallic minerals such as copper and iron, fuels such as coal, and, finally, emeralds. In the country of Colombia, salt was valuable to the mining industry because of its nonmetallic properties, use in food and industry, and health benefits. Even today, there are five important salt mines in Colombia, two of which are located in the department of Cundinamarca, 50 km from the capital city of Bogotá, and both produce salt from rock salt exploitation.

The excavation of four principal caverns started in Zipaquirá, Cundinamarca in 1816, and thanks to the invention of Alexander von Humboldt in 1801, the production of grain salt became more reliable. These four caverns were built in 1816, 1834, 1855, and 1876, respectively, and by 1881 the annual salt production of this mine was 8,400,000 kg. In October 1950, encouraged by the beliefs and faith of the mineworkers, the construction of a cathedral at a depth of 180 m began in the caverns dug by the Muisca two centuries ago. Fig. 1.7 shows

Figure 1.7 Passage inside the salt mine, and one of the caverns used for religious purposes in Zipaquira, Colombia.

the salt mine of Zipaquirá, which is also well known as the Salt Cathedral and recognized as Colombia's First Wonder.

1.3 WATER TUNNELS

A Chinese proverb roughly translated as "the water that bears the boat is the same that swallows it" suggests that water can be beneficial or detrimental depending on the situation.

Tunnels have always played an important role in the supply of water. Areas where the soil lacks sufficient rainfall are not fertile, making irrigation schemes involving tunnels necessary. Though water is essential for the survival of flora and fauna, it can also be destructive when in excess. Tunnels are therefore useful not only to transport water to where it is needed but also as irrigation to prevent excess of water in certain areas.

1.3.1 Tunnels to Supply Water

The famous ancient trade route known as Silk Road passed through the Taklamakan desert. To cope with its harsh climate, water management systems were crucial to supply drinking water and irrigation for caravans. A qanat (in Arabic) or karez (in Persian) is a water distribution channel with both underground (Fig. 1.8) and aboveground sections enabling the transport of water.

Water management systems, as shown below, use gravity to transport water from higher place to lower place with an arrangement of wells (Fig. 1.8). In spring and summer with melting snow and rain, large amounts of water pour

Figure 1.8 Photo of the inside of a karez (Winny, 2009).

down from mountains to and through the deserted soils and depressions. Locals built karez tunnels using ground slopes to cope with this (Hansen, n.d.).

The deepest known qanat is located in Gonabad in Iran. Built 2700 years ago, it is still used to provide water for both agricultural and drinking purposes for around 40,000 people (HCC, 2014).

1.3.2 Tunnels to Ensure Drainage

Drainage is critical in cities at risk of flood. Adequate systems thus need to be developed to limit surface waters (Fig. 1.9). A practice started by the Ancient Romans, drainage tunnels and sewer tunnels can be combined to do so (Fig. 1.10).

Events that occurred in Beijing and Ganzhou enable us to understand the paramount importance of drainage systems. On July 21, 2012 severe rainfall in Beijing led to widespread flooding across the Chinese capital, causing 79 deaths and $2 billion in damage (Zhang et al., 2013). Ganzhou, a smaller city in Jiangxi province, has also experienced similar rainfall on many occasions, but it has not led to such catastrophic flooding thanks to its ancient integrated drainage system.

Ganzhou is located between two rivers (Fig. 1.11). In order to control flooding and stormwater discharge, an efficient water management system was developed during the Northern Song dynasty (over 800 years ago).

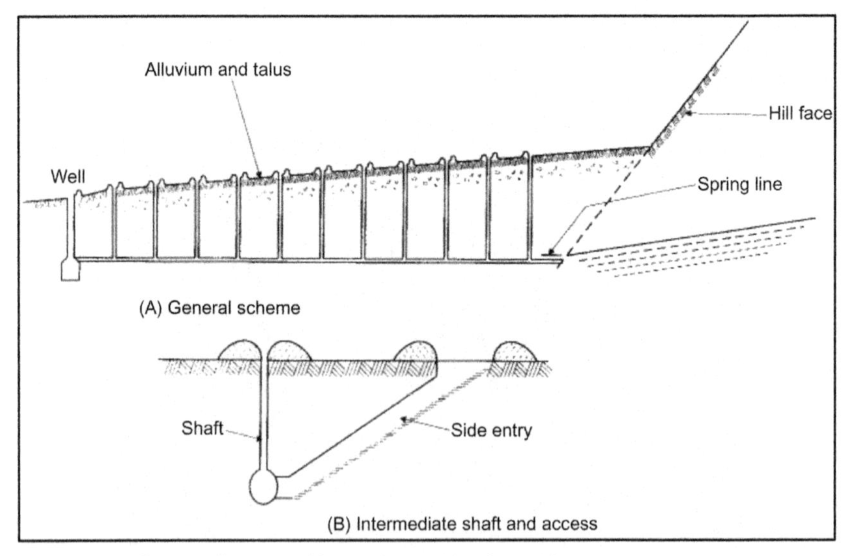

Figure 1.9 Scheme of a typical karez (Muir Wood, 2000).

Figure 1.10 A worker inspecting a 6th century drainage tunnel still in use today in Rome (Alvarez, 2006).

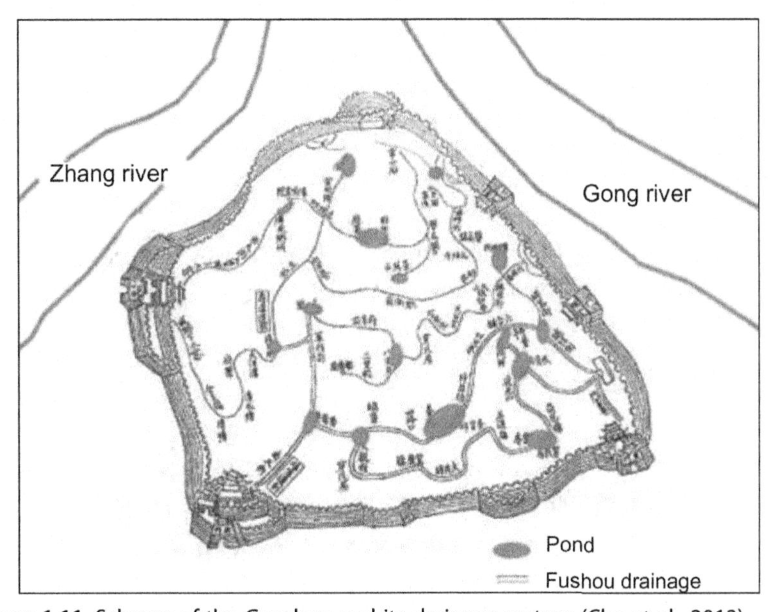

Figure 1.11 Scheme of the Ganzhou and its drainage system (Che et al., 2013).

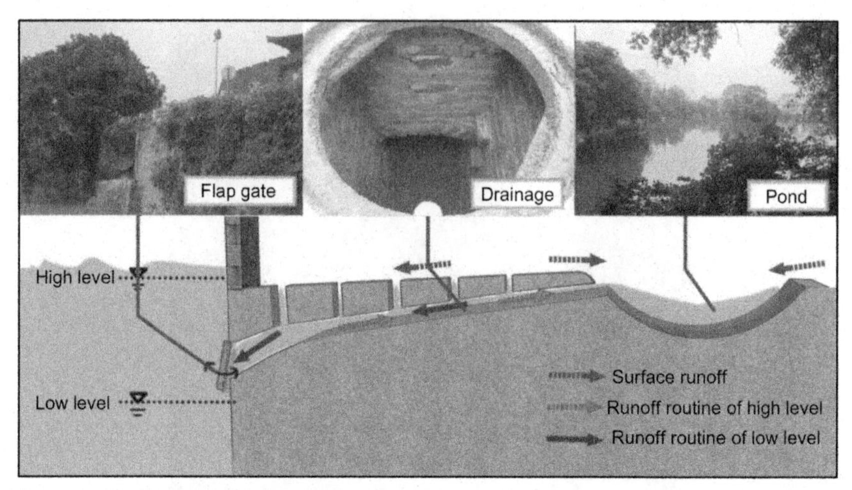

Figure 1.12 Diagram of the drainage system with the corresponding images (Che et al., 2013).

The Fushou drainage system was designed based on the layout and topography of the area. Its tunnels direct the flow of water into rivers and ponds depending on the quantity of water (Fig. 1.12). It is controlled through an ingenious system based on flap gates that open or close depending on the water level (Che, Qiao, & Wang, 2013).

1.4 UNDERGROUND POWER STATIONS

Underground power stations are hydropower stations that harness the energy of water poured from high reservoirs down through tunnels into a generating hall. Most large dams use their reservoirs in this way to generate electrical power. For example, the Three Gorges Dam project is the largest hydroelectric plant in the world with six underground generators. It is located in China in Hubei province and sits on Asia's longest river, the Yangtze.

Another example of the use of underground space to produce electricity is the Jinping II hydropower project in south-west China. But this hydropower plant uses underground space differently. It diverts water through a complex series of drainage tunnels as well as access and tailrace ones. Spanning almost 17 km (Fig. 1.13), it is the largest hydropower complex in the world (Fig. 1.14).

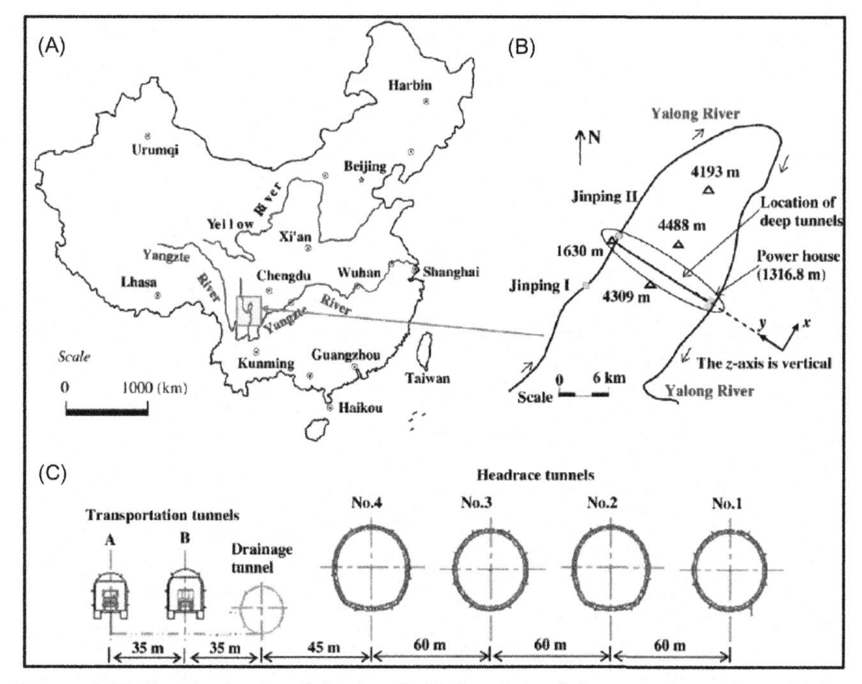

Figure 1.13 Sketched map of Jinping II: (A) location of the project in China, (B) layout of the hydropower project across the Yalong River, and (C) configuration of seven tunnels (Li et al., 2012).

Figure 1.14 Three-dimensional image of the Jinping II underground powerhouse (Wu, Shen, & Wang, 2010).

1.5 TRANSPORTATION TUNNELS

Between 2160 and 2180 BC, the Babylonians excavated a tunnel beneath the Euphrates river. It was allegedly used as an underpass for pedestrians and chariots, and according to archaeologists, was the first underwater transportation tunnel built (Browne, 1990).

Tunnel construction for transportation has increased in recent years. This is in part because of improvements in the tunneling industry that have enabled the construction of longer tunnels and through more challenging terrain. The growing need for more developed transportation infrastructures has also contributed to this increase. Many tunneling projects are therefore still underway today or are still being planned.

1.5.1 Railway Tunnels

Trains as a means of transportation have transformed society. The success of railway tunnels both for public transportation and for goods has led to their development and implementation across the world. The birth of the steam engine in the 19th century is undoubtedly the greatest contributor to the development of railways and consequently railway tunnel construction (Kjønø, 2017).

The first railway in China was built at the end of the 19th century, long after those in the western world. The first railway tunnel was the Shiqiuling tunnel, which passed through a narrow gauge in the Taiwan Province. China now has the world's largest high-speed railway network, most of which was built using conventional tunneling methods. However, with the development of tunneling technology, mechanized tunneling will soon be widespread. Tunnel engineers have come to realize that the mechanized tunneling method is not only faster but also less impactful to the environment, emitting fewer greenhouse gases.

The Qinghai-Tibet railway is a good example of a recent and ambitious railway tunnel project. It is the highest in the world and opened for transport service in 2006. It is a great achievement in the history of railway development, as the world's highest and longest plateau railway. The length of the railway is 1956 km and the average elevation is about 4500 m (Railway-Technology.com [RT], n.d.).

Due to the risks of high altitude and permafrost along the route, its construction was a challenge. It passes through numerous tunnels, including the world's longest plateau tunnel (Fig. 1.15) and most elevated tunnel

Figure 1.15 The entrance of the world's longest plateau tunnel, the 1686-m long Kunlun Mountain tunnel (Wong, 2015).

(RT, n.d.). Another tunnel, the 33-km-long Guanjian tunnel, was awarded the International Tunneling and Underground (ITA) award in 2006. The costs to build it exceeded 500 million euros.

1.5.2 Road Tunnels

Road and highway tunnels are used by automobiles and occasionally pedestrians and bicycles. Highway construction affects the environment considerably, and many new roads are being built and existing ones widened. Burying them reduces their environmental impact by reducing noise, congestion, and pollution. Another motivation for burying highways is that it frees the surface space for residents in urban environments. It can therefore improve living conditions by providing more open or recreational space.

Norway is a prime example of a road tunnel-building nation. The cost of its tunnels has been kept down by building most of them without linings, by using smooth blasting, and by taking advantage of suitable ground by carefully choosing routes. Due to its topography, Norway is covered in tunnels and has become an expert in rock tunneling technologies. Lærdal tunnel in Norway is a great engineering feat and serves as a model for tunnel engineers worldwide (BBC News, 2002),

Figure 1.16 Lærdal tunnel in Norway, the longest road tunnel in the world (BBC News, 2002).

with its state-of-the-art ventilation system and security-oriented lighting design (Fig. 1.16).

Inaugurated in 2007, the Zhongnanshan tunnel in China is the longest two-track road tunnel in the world and the longest overall tunnel in Asia with a length of over 18 km.

1.5.3 Metro Tunnels for Cities

In most urban transportation networks of significant size, the metro is the main mode of transportation It has many undeniable advantages, such as the high number of passengers it can transport, low energy consumption, security, punctuality, speed, and low cost. The metro is thus rapidly becoming the most favored means of transportation in cities across the world (Fig. 1.17). In the first 15 years of the 21st century, metro infrastructure worldwide grew by 40% (UITP, 2015).

As the roads of many cities are becoming more and more congested by traffic, the metro is becoming even more crucial. China is a good example of this growth of metro networks. From 2010 to 2014, the growth rate of urban metro lines in China was 97.3%. By 2020 their total length will reach 6000 km, spanning across 50 Chinese cities (Li, 2016). This demonstrates that although metro line construction began 100 years later in China, it now has three of the world's top 10 longest lines (Fig. 1.18).

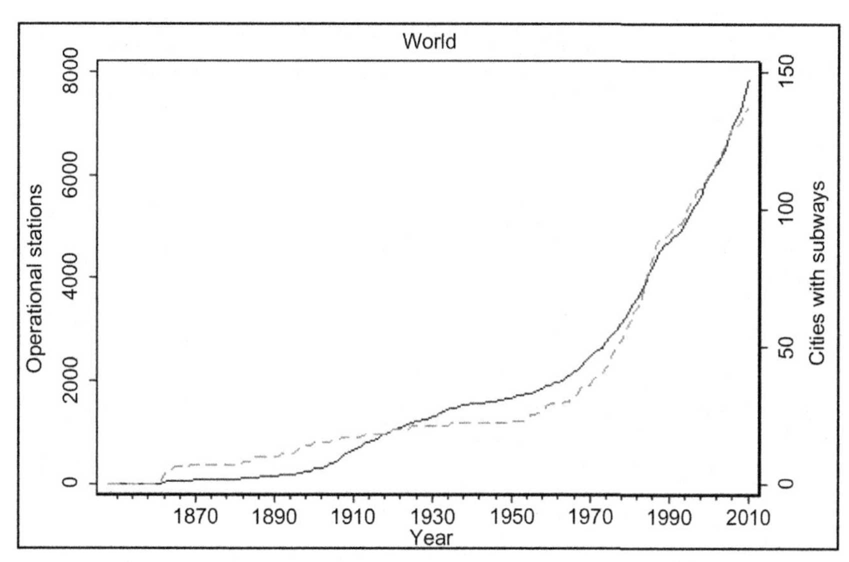

Figure 1.17 Growth of subway systems: the dashed curve represents the number of cities with subway systems, whereas the solid line represents the number of operational stations (Gonzalez-Navarro & Turner, 2016).

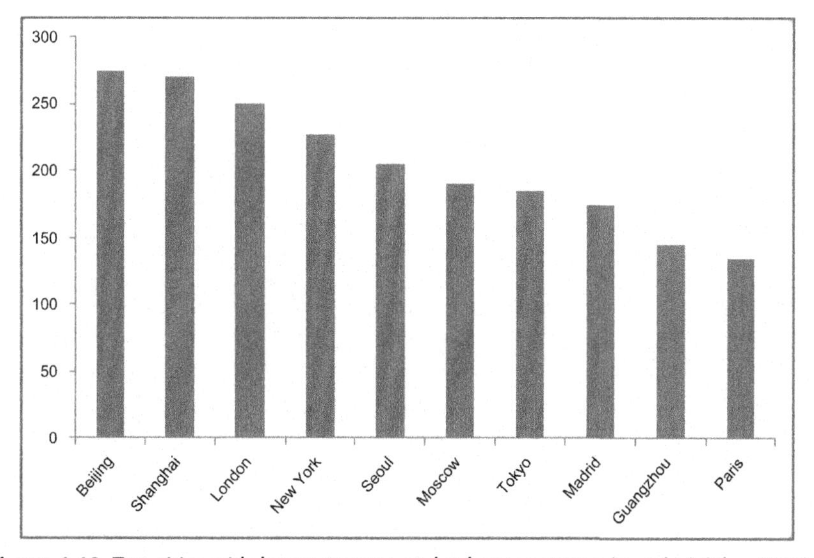

Figure 1.18 Top cities with longest metro and subway systems (in miles) (Zhu, 2013).

1.6 UNDERGROUND RECREATIONAL FACILITIES

There are several advantages of having underground recreational facilities. Being recreational, such spaces are only occupied by people for short periods of time. This means direct natural sunlight is not essential. Modern-day ventilation systems can provide clean and eventually filtered air from outside to meet the requirements of the space (Goel, Singh, & Zhao, 2012). Underground recreational space can also be located in urban environments, such as the workplace or housing areas. This removes travel time to and from the underground recreational spaces for people that live or work just above them. In the Chinese cities of Hangzhou and Shanghai, this has already been implemented successfully (Goel et al., 2012).

In Finland a wide range of facilities have also been built underground. In addition to cumbersome and noisy water plants, recreational spaces such as the swimming pool in Itäkeskus (Fig. 1.19) have also been built beneath the surface. This recreational facility is a good example of wise planning of the use of underground space. Along with the ability to host 1000 swimmers at a time, the space can also be converted into an emergency shelter to accommodate 3800 people (Vähäaho, 2014).

Figure 1.19 The underground swimming pool in Itäkeskus (Makkonen, 2014). *From Makkonen, E. (2014). Underground swimming pool [Photograph]. In Vähäaho, I. (2014). Underground space planning in Helsinki. Journal of rock mechanics and geotechnical engineering, 6(5), 387–98, 'Underground swimming pool in Itäkeskus, which can accommodate 1000 customers at a time and can be converted into an emergency shelter for 3800 people if necessary (Photo: Erkki Makkonen)'.*

1.7 UNDERGROUND SPACE FOR THE FUTURE

Underground space for the future will be based on current research and development. In addition to the examples just discussed, future underground space will also attempt to innovatively use underground urban space in unprecedented ways. In China, for instance, Shanghai's government has decided to explore the underground space in the following six directions:

1. Strategic study of Shanghai underground space exploitation
2. Future underground construction technology and equipment study
3. Deep underground space planning theory and design study
4. Low-carbon underground space exploration technology study
5. Safe operation and digitalization study of underground construction's full life cycle
6. Study on risk management and long-term evaluation and experiments

1.7.1 Underground Space in Cities

Urban environments can function more efficiently using multipurpose underground space to alleviate the pressure on the surface. As previously detailed, metro lines in cities across the world already alleviate the transportation networks aboveground. Other uses of underground space include the Earthscraper project proposed by Mexican architects BNKR Arquitectura. It is a 65-story inverted skyscraper that plunges 300 m below ground (Dvice.com [DV], 2011).

Another innovative project is the underground cemetery in Jerusalem. To meet the high demand for burial space without expanding existing cemeteries on the surface, the city of Jerusalem decided to expand one depth-wise. Beneath its main cemetery, tunneling teams are working on 45-m-deep structures that will eventually add another 22,000 burial spots (Liebermann, 2016). This solution saves surface space while maintaining proximity to the cemetery for visitors. The project is a good example of how a centuries-old burial method (catacombs) is now being implemented with modern technologies.

1.7.2 Future Tunnels

Great mountains and deep waters separate and isolate land, posing a barrier to the movement of freight and people to and from different locations. The significant economic and technological growth of the 21st century has led and is leading to the construction of many worldwide transportation networks, with the aim of providing safer, faster, and more cost-effective mass transportation.

One of many intercontinental projects is the Gibraltar tunnel that would link Europe and Africa. At around 40 km long, it would lie beneath the strait of Gibraltar between Morocco and Spain. As a twin-rail tunnel, it is often compared to The Channel Tunnel that links France and England. But the project faces multiple challenges. Namely, unfavorable geological conditions and a sea depth of over 300 m in certain areas. The feasibility of connecting the railway networks of Morocco and Spain through an undersea tunnel is therefore still being studied (Hamilton, 2007).

1.7.3 Energy Supply Research

Some vast projects have already been planned for the coming years. Tunnels play a role in the supply of energy in many ways, namely with water flow through them (hydroelectricity), in oil and gas pipelines, etc. One project in Norway envisions exploiting oilfields off the coast. Subsea tunnels would link onshore facilities to oil drilling and operational facilities located 30 km from the shore (Grøv, Nilsen, & Bruland, 2013).

China also seeks to exploit its reserves of shale gas. Wells several thousand meters deep need to be dug to reach it, and machines have been built to do this while minimizing environmental impact (GeoResources, 2014). For example, a shaft-boring machine designed by Herrenknecht can safely, economically, and rapidly reach deep deposits for mining. It produces blind shafts at a depth of up to 2000 m (Herrenknecht, n.d.).

1.7.4 Submerged Floating Tunnels

Developing tunnels on land, under or through the sea, pose different challenges. Existing technologies include immersed tubes and submerged tunnels (Fig. 1.20). Depending on the constraints and requirements of a project, each can be adapted (Fig. 1.21). Submerged floating tunnels lie beneath shipping levels and are tethered to the seabed. They can be used to cross great widths of water such as rivers, lakes, and bays. Submerged floating tunnels are a combination of tunnels and off-shore structures (Mai & Guan, 2007) and have numerous advantages (Wallis, 2010). They can be shorter than other alternatives as well as cheaper and faster to build, as fewer construction materials are needed.

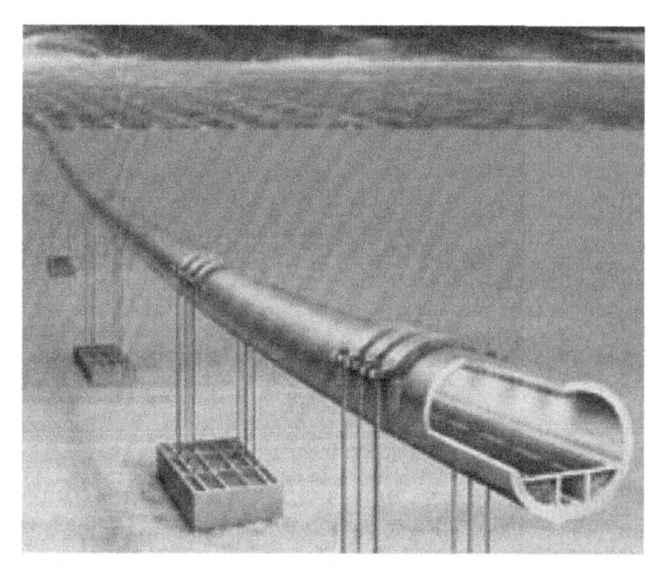

Figure 1.20 Submerged floating tunnel (Wallis, 2010).

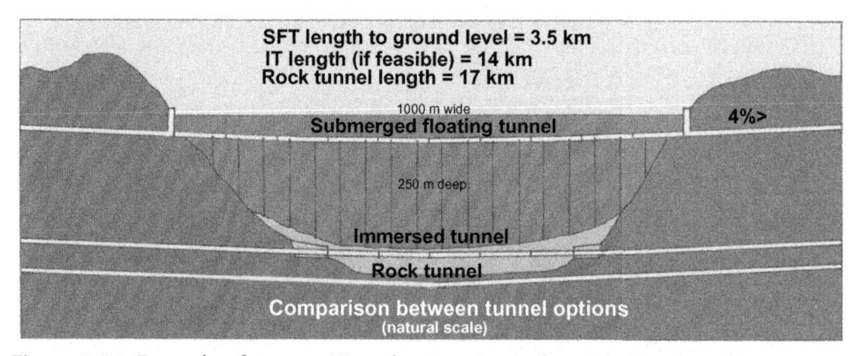

Figure 1.21 Example of a comparison between tunnel options (Wallis, 2010).

1.7.5 Moon Caverns

According to the Japan Aerospace Exploration Agency (JAXA), the Moon will soon become a base for mankind's activities. Since 2004, Japanese scientists have discovered multiple caves beneath the Moon's surface called lava tubes. The latest lava tube discovered in 2017 is a cave 50 km long and 100 m wide. It appears to be structurally sound and its rocks may contain ice or water deposits that could be turned into fuel, according to data sent back by the orbiter to the Marius Hills Hole.

Fig. 1.22 shows the echoes from the subsurface; the first echo peak (red point; normalized to 0 dB) is from the surface, and the second one (blue dot) must be from a subsurface boundary. It is noted that prior to the second echo peak, the received echo power precipitously decreased with time to a noise level of 28.1 dB (green dot). This echo pattern with two peaks and the substantial echo decrease between them implies the existence of a cave, such as an underlying lava tube. After the second echo peak, the received echo power decreased (orange dot). The purple diamond marks the third echo peak.

Due to their stable thermal conditions and potential to protect people and instruments from micrometeorites and cosmic ray radiation, these caverns could be used as a base for astronauts and their equipment as well as other human explorers. Scientists believe that this type of cave could also become a base for a future lunar human colony.

As Stephen Hawking noted, "we will not survive another 1000 years without escaping beyond our fragile planet" (Holley, 2017). As space and resources on Earth are limited, mankind must consider extraterrestrial exploration. The Moon may become its first space colony. But due to its severe environment, a moon colony would have to be built underground.

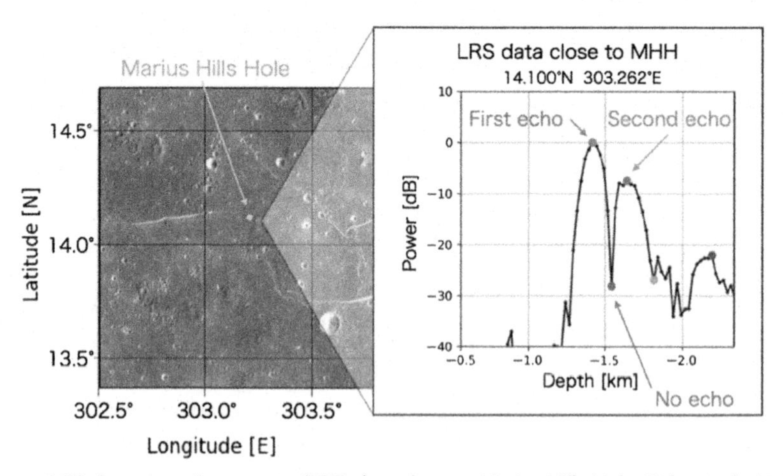

Figure 1.22 Laser ranging system (LRS) data close to Marius Hills Hole (Kaku et al., 2017).

1.8 QUESTIONS

1.1. What are the purposes of the use of underground space?

1.2. Why do we need to go underground and what are the benefits of underground structures?

1.3. Can underground spaces help solve typical urban problems such as traffic congestion, pollution, and noise? Explain.

1.4. Do you think underground structures are more resilient to natural disasters than structures at the surface. Why?

1.5. Can you think of other uses of underground space besides the ones described in this chapter?

REFERENCES

Alvarez L.S. (2006, July). *A tunnel inspection in Roma* [Photograph]. Retrieved from <http://ngm.nationalgeographic.com/features/world/europe/italy/romeruins-text/3>

AncientCraft (n.d.). *Grimes graves* [Drawing]. Retrieved from <http://www.ancientcraft.co.uk/Places/grimes_graves.html>

Bhalodiya, R. (2016a, November 28). *The top 10 longest road tunnels all over the world.* Retrieved from <http://businessmensedition.com/top-10-longest-road-tunnels-world/>

Bhalodiya, R. (2016b, November 29). *The top 10 longest rail tunnels of the world.* Retrieved from <http://businessmensedition.com/worlds-top-10-longest-rail-tunnels/>

BBC News. (2002, April 24). Light at the end of the tunnel. *BBC News*. Retrieved from <http://news.bbc.co.uk/2/hi/europe/1945581.stm>

Browne, W.M. (1990, December 2). Tunnel drilling, old as Babylon, now becomes safer. *The New York Times*. Retrieved from <http://www.nytimes.com/1990/12/02/world/tunnel-drilling-old-as-babylon-now-becomes-safer.html>

Cascone, S. (2014, September 30). The getty helps save China's Mogao Grottoes from tourists. *Artnet News*. Retrieved from <https://news.artnet.com/exhibitions/the-getty-helps-save-chinas-mogao-grottoes-from-tourists-118469>

Che, W., Qiao, M. X., & Wang, S. S. (2013). Enlightenment from ancient Chinese urban and rural stormwater management practices. *Water Science & Technology, 67*(7), 1474–1480.

Crystalinks. (n.d.). *Prehistoric mining*. Retrieved from <http://www.crystalinks.com/prehistoric_mining.html>

Daily Mail Reporter. (2011, August 8). Going underground: The massive European network of stone age tunnels that weaves from Scotland to Turkey. *Mail Online.* Retrieved from <http://www.dailymail.co.uk/sciencetech/article-2022322/The-massive-European-network-Stone-Age-tunnels-weaves-Scotland-Turkey.html>

Dvice.com. (2011, October 3). *The Earthscraper: An inverted pyramid 65 storeys deep.* Retrieved from <http://www.redicecreations.com/article.php?id = 17039>

GeoResources. (2014, May 9). Efficient shale gas development in China with German mobile deep drilling rig. *GeoResources*. Retrieved from <http://www.georesources.net/index.php/global-news/efficient-shale-gas-development-in-china-with-german-mobile-deep-drilling-rig>

Goel, R. K., Singh, B., & Zhao, J. (2012). *Underground infrastructures: Planning, design and construction* (1st ed.). Waltham, MA: Butterworth-Heinemann.

Gonzalez-Navarro, M., & Turner, A. M. (2016). Subways and urban growth: Evidence from Earth, April *SERC Discussion Paper, 195*.

Gorski, H. (2006). *Hal Saflieni Hypogeum* [Photograph]. Retrieved from <http://whc.unesco.org/en/list/130>

Grøv, E., Nilsen, B., & Bruland, A. (2013, November). Subsea tunnels for oilfield development. *TunnelTalk*. Retrieved from <https://www.tunneltalk.com/TunnelTECH-Nov2013-Development-of-subsea-tunnel-access-to-offshore-oil-fields.php>

Hamilton, R. (2007, March 13). Africa and Europe set for tunnel link. *BBC News*. Retrieved from <http://news.bbc.co.uk/2/hi/africa/6442697.stm>

Hansen, R.D. (n.d.). Karez (Qanats) of Turpan, China. *WaterHistory.org*. Retrieved from <http://www.waterhistory.org/histories/turpan/turpan.pdf>

HCC. (2014, November). *Tunnelling: Coming through the ages*. 5 pp. Retrieved from <http://www.hccindia.com/whitepaper_pdf/2014/tunneling-november-2014.pdf>

Herrenknecht (n.d.). *Shaft Boring Machine (SBM) – Safe and quick construction of blind shafts down to depths of 2,000 meters*. Retrieved from <https://www.herrenknecht.com/en/products/core-products/mining/shaft-boring-machine-sbm.html>

Holley, P. (2017). Stephen Hawking now says humanity has only about 100 years to escape earth. *Chicago Tribune*. Retrieved from http://www.chicagotribune.com/news/nationworld/science/ct-stephen-hawking-escape-earth-20170505-story.html.

Kaku, T., Haruyama, J., Miyake, W., Kumamoto, A., Ishiyama, K., Nishibori, T., … Howell, K. C. (2017). *Detection of Intact Lava Tubes at Marius Hills on the Moon by SELENE (Kaguya) Lunar Radar Sounder. Geophisical Research Letters*. AGU Publications. Retrieved from http://www.isas.jaxa.jp/en/topics/001159.html.

Kjønø, E. B. (2017). *4000 years of tunnelling. In Horn International. Mining & Tunnelling* (4th ed, pp. 8–13)Oslo, Norway: Horn International AS.

Li, K.M. (2016, March). *A brief introduction to China's PPP application in transport and logistics sectors* [PowerPoint slides]. Retrieved from <https://www.unece.org/fileadmin/DAM/ceci/documents/2016/PPP/Forum_PPP-SDGs/Presentations/Kaimeng_LI-UNECE_PPP_Forum_March_2016_A_Brief_Introduction_to_China%E2%80%99s_PPP_Application_in_Transport_and_Logistics_Sectors.pdf>

Li, S. J., Feng, X. T., Li, Z. H., Chen, B. R., Zhang, C. Q., & Zhou, H. (2012). In situ monitoring of rockburst nucleation and evolution in the deeply buried tunnels of Jinping II hydropower station. *Engineering Geology, 137–138*, 85–96.

Liebermann, O. (2016, September 25). Going inside Jerusalem's underground city of the dead. *CNN*. Retrieved from <http://edition.cnn.com/2016/09/25/middleeast/jerusalem-underground-city-for-the-dead/index.html>

Mai, J. T., & Guan, B. S. (2007). Submerged floating tunnel. *Tunnel Construction*, *27*(5), 20–23.

Makkonen, E. (2014). Underground swimming pool [Photograph]. In Vähäaho (2014).

Muir Wood, A. (2000). *Tunneling: Management by design*. London, England: E & FN SPON.

Railway-Technology.com. (n.d.). *Qinghai-Tibet heavy rail line, China*. Retrieved from <http://www.railway-technology.com/projects/china-tibet/>

Ring, T., Salkin, R. M., & La Boda, S. (1995). *Malta, Internationaldictionary of historic places* (Vol. 3, pp. 411–414). Chicago, IL: Fitzroy Dearborn.

Swaziland National Trust Commission. (n.d.). *Cultural resources: Malolotja Archaeology, Lion Cavern*. Retrieved from <http://www.sntc.org.sz/cultural/malarch.asp>

taQpets (2013). *Cave dwellings not far from Beijing* [Photograph]. Retrieved from <http://www.solaripedia.com/13/408/6380/cave_dwellings_in_china.html>

Thacker, S. (2016, February 28). The Summit tunnel 175 years on: The miracle of engineering which survived a devastating fire. *Manchester Evening News*. Retrieved from <http://www.manchestereveningnews.co.uk/news/nostalgia/summit-tunnel-175-years-on-10959530>

UITP. (2015, October). *World metro figures – statistics brief.* Brussels, Belgium.

UNESCO. (n.d.). *Mogao caves.* Retrieved from <http://whc.unesco.org/en/list/440>

Vähäaho, I. (2014). Underground space planning in Helsinki. *Journal of Rock Mechanics and Geotechnical Engineering, 6*(5), 387–398.

Wallis, S. (2010, January). Links across the waters. *Strait crossings conference report.* Retrieved from <https://www.tunneltalk.com/Strait-Crossings-Jan10-Conference-report.php>

Winny (2009, November 9). *Inside of a karez* [Photograph]. Retrieved from <http://www.absolutechinatours.com/Turpan-attractions/Karez-5321.html>

Wong, D. (2015, September 16). *Top 10 statistics you do not know about Tibet railway.* Retrieved from <http://www.tibettravel.org/qinghai-tibet-railway/facts.html>

Wu, S. Y., Shen, M. B., & Wang, J. (2010). Jinping hydropower project: Main technical issues on engineering geology and rock mechanics. *Bulletin of Engineering Geology and the Environment, 69*(3), 325–332.

Yoon, H. (1990). Loess cave-dwelling in Shaanxi Province, China. *GeoJournal, 21*(1), 95–102.

Zhang, D. L., Lin, Y. H., Zhao, P., Yu, X. D., Wang, S. Q., Kang, H. W., & Ding, Y. H. (2013). The Beijing extreme rainfall of 21 July 2012: "Right results" but for wrong reasons. *Geophysical Research Letters, 40,* 1426–1431.

Zhu, W.J. (2013, May 2) China's subway construction frenzy. *The world of Chinese.* Retrieved from <http://www.theworldofchinese.com/2013/05/chinas-subway-construction-frenzy/>

FURTHER READING

Freudenrich, C. (2007). How pyramids work. *How Stuff Works.* Retrieved from <http://science.howstuffworks.com/engineering/structural/pyramid2.htm>

NAT. (2013). *Underground engineering for sustainable urban development.* Washington, DC: The National Academic Press978-0-309-27824-9. Retrieved from http://nap.edu/14670.

CHAPTER 2

Planning the Use of Subsurface Space

Contents

The underground examples discussed in Chapter 1, History of Subsurface Development, show that utilization of underground space is crucial for the development and sustainability of modern cities around the world. There is high demand for infrastructure as urbanized populations continue to grow rapidly (Fig. 2.1) and existing infrastructures fail. To cope with this demand and rapid growth, the exploitation of underground space is essential. This also applies to smaller urban environments such as small and medium-sized towns, where underground space is still needed for municipal pipeline infrastructures, for example.

Underground Engineering
DOI: https://doi.org/10.1016/B978-0-12-812702-5.00002-5

Figure 2.1 Traffic congestion (left) and aboveground crowded environment (right) in Shanghai, China.

Helsinki is a prime example of proper underground planning. The entire municipal area has an underground plan that takes both sustainability and esthetics into account as well as longevity in the development of its underground urban space (Vähäaho, 2014).

In Japan, some cities have underground traffic that has developed to such an extent that its congestion is similar to that aboveground (ITA Working Group no.4 [ITA WG4], 2000).

Underground projects have profound influences on society and the environment, and a lack of systematic planning can lead to detrimental consequences that may or may not be reversible. According to a Chinese idiom, painting bamboo is easier if the finished and painted bamboo is already in one's mind.

Although the planning of underground projects may appear tedious, the success of these projects is greater when it is properly defined. Proper planning prevents late modifications that often lead to added costs and delays. In this chapter, the basis and general features of underground project planning will be introduced.

2.1 ECONOMIC BENEFITS

At first glance, underground projects may seem expensive for what they provide. Though they require large initial investments that mean they are almost always more expensive than those aboveground (Table 2.1), there are often extra benefits. Occupying the underground saves space

Table 2.1 Cost comparison between different alternatives in the case of an urban mass transit system

Construction	Initial capital cost ratio	Total lifecycle cost ratio
Underground	3.2–4.5	0.4–0.5
Surface	1	1
Elevated	2–2.8	1.4

Data from ITA working group no.20., 2012, April. ITA report no.011, Report on underground solutions for urban problems and Tunnels and Tunnelling, 2005. Retrieved from https://tunnelsonline.info.

Table 2.2 Annual cost of congestion (Fenalco, 2014; INRIX/Cebr, 2014)

Country	Annual losses in USD
France	469,000 million
Germany	691,000 million
United States	2.8 billion
Colombia	3.9 billion

aboveground. In cities, this leads to large savings in terms of compensation and claims (ITA Committee on Underground Space, 2009). Underground solutions also have long-term benefits such as lower lifecycle costs due to their high durability compared to aboveground projects (ITA Working Group no.20, 2012; Tunnels and Tunneling, 2005).

Congestion in cities is also a major problem as it reduces the functionality of road networks. Travel take longer, leading to large economics losses. According to studies on the future impacts of traffic jams in the United Kingdom, Germany, France, and the United States (Fenalco, 2014; INRIX/Cebr, 2014), by 2030 congestion will cost an average family living in London more than 4000 USD a year in lost time. Even in South American countries like Brazil and Colombia, the estimated annual cost of congestion is alarming. Table 2.2 shows such results based on direct and indirect costs. Direct costs relate to the value of fuel and the time wasted rather than being productive at work, and indirect costs relate to higher freighting and business fees from company vehicles idling in traffic, which are passed on as additional costs to consumers.

City congestion can be reduced by promoting alternative means of transportation, such as public buses, trams, and subways. But the costs must be considered on a life-cost basis (Fig. 2.2) rather than on the cost of the initial investment.

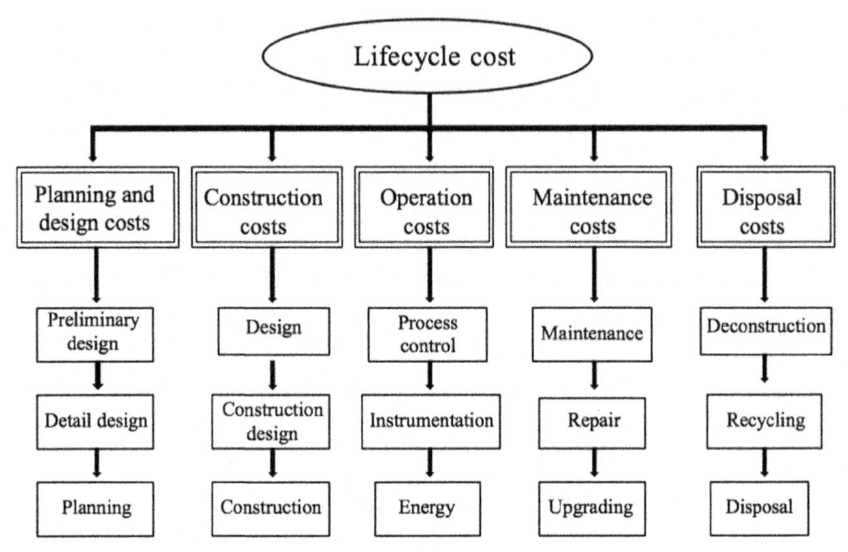

Figure 2.2 Lifecycle cost consideration.

2.2 TYPES OF UNDERGROUND SPACES

As discussed in Chapter 1, History of Subsurface Development, underground space can be exploited for many reasons. The most common examples include:

- transportation (metros, roads, railways, etc.);
- distribution (power, water, sewage, etc.);
- storage (water, goods, energy, car parks, etc.);
- recreational and commercial (sports grounds, zoos, shops, etc.);
- educational (libraries, test spaces, etc.);
- industrial (factories, workshops, offices, etc.);
- for defense (control centers, military installations, etc.); and
- cemeteries.

Some facilities must be built underground, such as sewage and certain water networks. But underground space is usually a competitive alternative. In the planning of such projects, one should therefore be aware of the merits of underground solutions. The constraints imposed on an underground project depend on its intended application. For example, rail and metro tunnels are less flexible in alignment and slope than those intended for road traffic.

2.3 DEPTH IN UNDERGROUND PLANNING

Underground planning is different from traditional urban planning as it must consider a third dimension—depth. Different levels beneath the surface serve for different applications. Fig. 2.3 illustrates a common underground structure arrangement. The same figure also shows how crowded underground space can become. In recent years in Japan, new underground construction work is being carried out at greater depths (ITA WG4, 2000), for applications such as rail, road, regulating ponds, etc., because of this.

Thus, depth considerations have to be defined when planning different structures, as explained in Section 2.4.1. In China and elsewhere, underground research is being carried out to gain a better understanding of the technical difficulties that exist when working at increasing depths for extended periods of time.

2.4 UNDERGROUND LEVEL PLANNING

2.4.1 Criteria

It is generally accepted that urban surfaces should be reserved for the most important functions, namely work, recreation, and housing. As cities

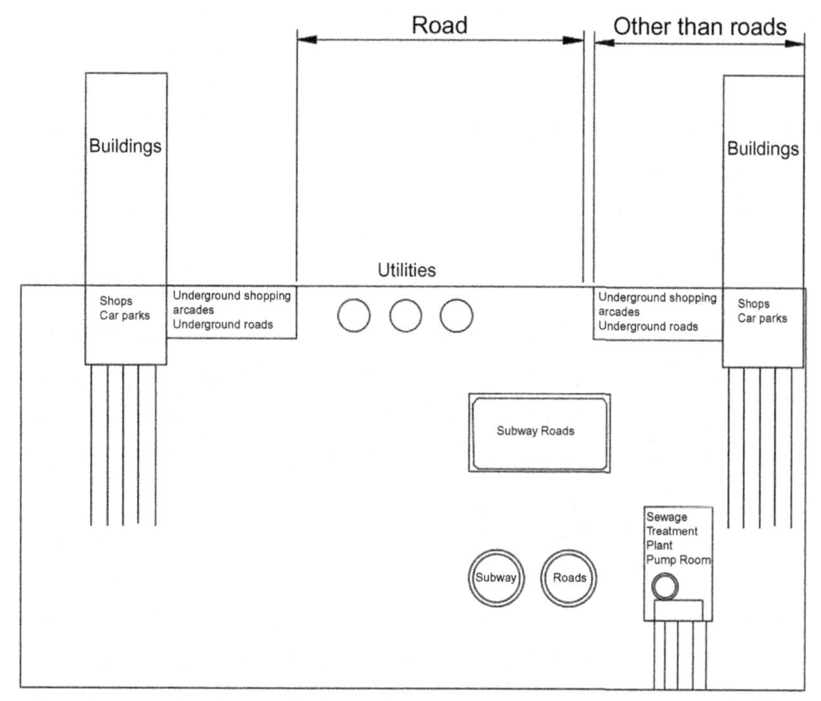

Figure 2.3 Uses of underground space in Japan in relation to land ownership (ITA WG4, 2000).

are becoming increasingly dense, people are moving recreational spaces underground, while residential buildings and full-time work facilities are still located aboveground.

If it is possible to integrate surface and subsurface use, the surface can be called level 0, or 0 m depth. Table 2.3 illustrates the approach taken in underground planning: the more frequently a facility is used, the shallower it lies in the ground. Ground-level and shallow constructions are

Table 2.3 Vertical arrangement for underground engineering projects

Location		Depth (m)	Usage
Under the road	Near surface	0 to −3	Utilities and subway stations
	Shallow layer	−3 to −15	Subways, underground passage and main utilities
	Medium layer	−15 to −40	Subways, underground railways, underground road tunnels, and underground logistic flow tunnels
	Deep layer	< −40	Underground water resource, special underground structures, and space for future development
Not directly under road	Shallow layer	0 to −15	Subways, underground hubs, underground streets, civil defenses, underground garage, underground reservoirs, underground transformer stations, and foundations of buildings
	Medium layer	−15 to −40	Underground logistic flow tunnels, garages for dangerous items, underground railways, underground road tunnels, and foundations of buildings
	Deep layer	> −40	Underground water resource, special underground structures, and space for future development

suitable for living. At deeper levels there are facilities that require trained personnel and at the deepest levels there are the unmanned ones.

Other criteria are also considered in the level planning of an underground space:

- The smaller the structure, the shallower it should be.
- Facilities used for public traffic should have higher priority than those for private ones.
- Convenience for pedestrians takes priority over that for drivers.

2.4.2 Considerations

Based on frequency of use, underground engineering projects can be divided into different categories (Ronka, Ritola, & Rauhala, 1998):

1. Underground space frequently used by the public.

   This includes commercial and recreational zones. The main planning consideration is to create a healthy and comfortable environment. Special attention must therefore be paid to lighting, ventilation, acoustics, and ease of orientation and movement.
2. Underground space used intensively by the public.

   This refers to underground traffic networks (namely metro lines) and parking lots. Planning should give priority to convenience and accessibility.
3. Underground space used only by a specific group of people.

   This refers to technical maintenance facilities such as sewage treatment plants, power plants, and storage spaces. As such facilities are rarely visited, they tend to be situated at the deep underground level.
4. Underground space seldom visited.

   This includes telecommunication cable tunnels and sewage and water supply tunnels. Like the previous category, this space can be built at high depth as it is seldom visited.

2.4.3 Planning Phases

Underground planning can be divided into three phases: preliminary, overall, and detailed planning (Table 2.4). Each phase has its own priorities and phases may overlap.

2.4.4 Sustainable and Integrated Planning

With the economic growth of many cities, their development requires the use of the subsurface space in addition to the surface space. This

Table 2.4 Objectives of different planning stages

Stage	Objectives
Preliminary planning	1. To indicate various needs and purposes 2. To present relevant basic data (current land use, geology, etc.) 3. To indicate planning goals 4. To provide reasons for choosing certain alternatives 5. To indicate social and environmental effects 6. To coordinate the interests of different stakeholders 7. To address specific problems of feasibility and economics
Overall planning	1. To outline unique resources and general features of subsurface space 2. To describe special subsurface uses 3. To sketch installations 4. To edit recommendations on future subsurface use 5. To plan the construction 6. To present geotechnical data
Detailed planning	1. To coordinate what is above and under the surface 2. To predict excavation conditions to a great extent 3. To define the highest standards in the interpretation of geological and geotechnical data 4. To specify the cost as there can be unexpected hazards (avoidable by prior high-quality studies, investigations, and assessments) 5. To combine a thorough understanding of theory and practice

growth has also affected how infrastructures themselves have developed. Combining aboveground and underground structures is crucial to the development of interconnected sustainable spaces for cities.

We all know that the underground can be used to separate or isolate hazardous materials such as raw sewage or high-voltage electrical lines from people and infrastructures on the surface. On the other hand, this separation means that protecting against physical hazards such as flooding, internal fire, and explosions is more challenging, especially as diverse underground infrastructure becomes more integrated with other underground and surface infrastructures. Therefore, the integration of surface and subsurface is crucial for safety.

A new age of the use of underground space integrated with surface facilities began as a solution for urban activities that could not be carried out at the surface alone. According to the World Population Prospect, by 2050 the countries that already have the largest populations will contribute the most to the projected increase of the world's population

(UN, 2014). Furthermore, most of them are developing countries that are already facing the challenges brought by this growth. As the population of a city increases, its urban development requires a reliable infrastructure that can meet the needs of the city. It is easy to see that the development of underground space has solved transportation issues in big cities, but in addition to free land, the underground offers varied and better solutions for things such as car parking, commerce and entertainment, water storage or wastewater treatment facilities, sewerage systems, hydropower stations, and nuclear waste storage as discussed in Section 2.2.

In the French city of Lyon, for example, updates was needed on the Croix-Rousse tunnel, due to changes in safety regulations and repeated neglect. To optimize investments and foster sustainable development, a safety tube was designed as an ecofriendly soft transport mode tunnel (Labrit, Chatard, Walet, & Dupont, 2012). It is a three-lane traffic tunnel: one double-track lane for cyclists, one for pedestrians, and the third for buses. Parallel to the road tunnel, its total length is 2 km (Fig. 2.4).

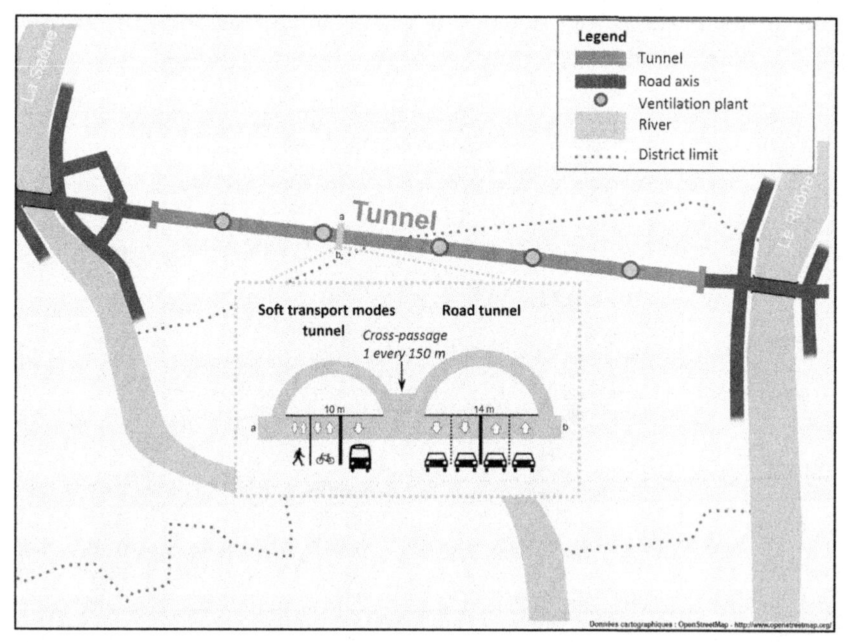

Figure 2.4 Overview of the Croix-Rousse tunnel. *Adapted and translated from Tibidibtibo (2013).* Tunnel de la Croix-Rousse – Lyon *[diagram]. Retrieved from https://commons.wikimedia.org/wiki/File:Tunnel_de_la_croix_rousse.svg*

2.5 FINANCIAL PLANNING

2.5.1 Principles of Financial Planning

Engineering projects require financial sustainability. In the private sector, this financial sustainability is to ensure profit, while in the public sector it is to provide a public service while minimizing costs. An inadequate financial plan may lead to poor preparation, inability to complete part of the contract, or even project failure.

In a typical underground engineering project, there are three stages in financial planning. These are:

1. Preparation—when costs are calculated
2. Construction—when most of the money is spent
3. Operation—when costs are recovered

Financial planning is not a definitive process as there are always risks associated with not being able to meet financial needs. Fig. 2.5 provides a clear visualization of cost and risk fluctuations throughout the different stages. The dashed line labeled "Committed investment" stands for the actual cost. This line has a gentle slope before the construction starts, indicating that only a small percentage of the overall cost has occurred (due to site investigation, design, etc.). The slope then increases sharply during construction.

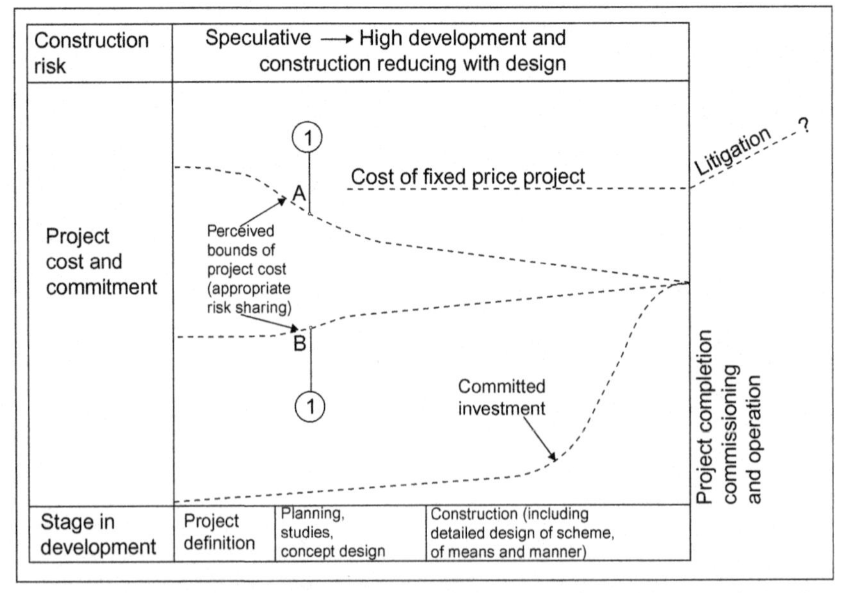

Figure 2.5 Changing perceptions of cost as a project develops (Muir Wood, 2000).

Fig. 2.5 shows that as a project proceeds and costs increase, risks are gradually reduced as knowledge of all the influencing factors improves. Note that the gap between the two dashed lines "Perceived bounds of project cost" (representing the risk) converges to zero. This is because as the project proceeds, risk is reduced from its initial speculative stage to zero upon completion. The vertical line connecting ①-① in Fig. 2.5 shows that the estimated project cost at the planning stage is considered a stochastic variable.

To illustrate how the perceived upper and lower boundaries are defined, the vertical line connecting ①-① is taken out (Fig. 2.6) to expand the information along this line. At any time, the anticipated project cost is a random variable, and the perceived upper and lower bounds defined in Fig. 2.5 are low and high estimates (Muir Wood, 2000).

2.5.2 Financial Mechanisms and Responsibilities

In underground engineering projects, it is generally accepted that it is a government's responsibility to develop and maintain the infrastructure. In most cases, a government may seek help from the private sector for expertise and/or other resources. Public–private partnerships (PPPs or P3s) were created for this purpose. This form of cooperation is advantageous as it combines the skills and resources of both the public and private sectors.

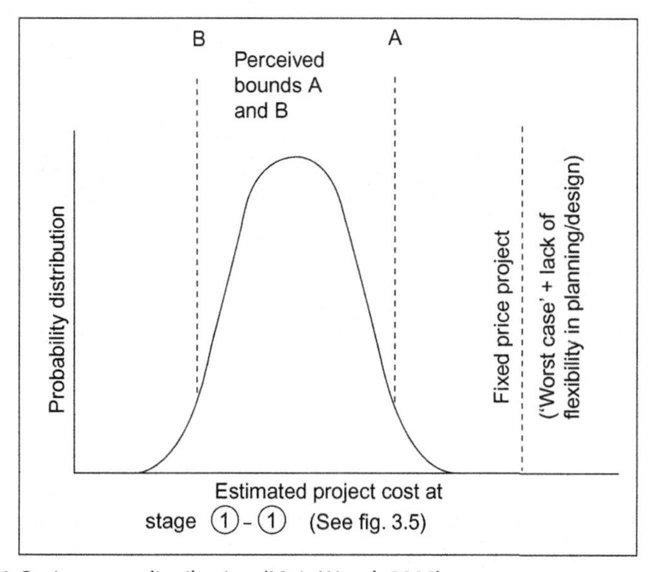

Figure 2.6 Project cost distribution (Muir Wood, 2000).

Risks and responsibilities are also shared, allowing governments to benefit from the expertise and efficiency of the private sector and to focus on policy, planning, and regulation. The private sector also benefits as its technologies and innovations are used, which means higher profit margins. PPP projects are funded in several ways:

1. project funding,
2. government funding, and
3. private funding from corporate or on-balance sheet finance.

Among these, project funding is the most common and efficient way of funding PPP projects (PPPIRC, 2016).

Fig. 2.7 shows the typical structure of project financing for a build, operate, and transfer project. The core of this structure is the creation of the project's company (special-purpose vehicle), which is created for a specific PPP project. It has the following features:

1. It has no previous business or record.
2. Its sole activity is to carry out the project through subcontracting (construction contracts and operation contracts).
3. For a new build project, there is no revenue during the construction stage and it only generates revenue once the project is in the operation phase (i.e., there are significant risks for all stakeholders during the construction phase).

In project financing, the project company can be considered a tool that links all stakeholders of the project. For example, suppose that a city

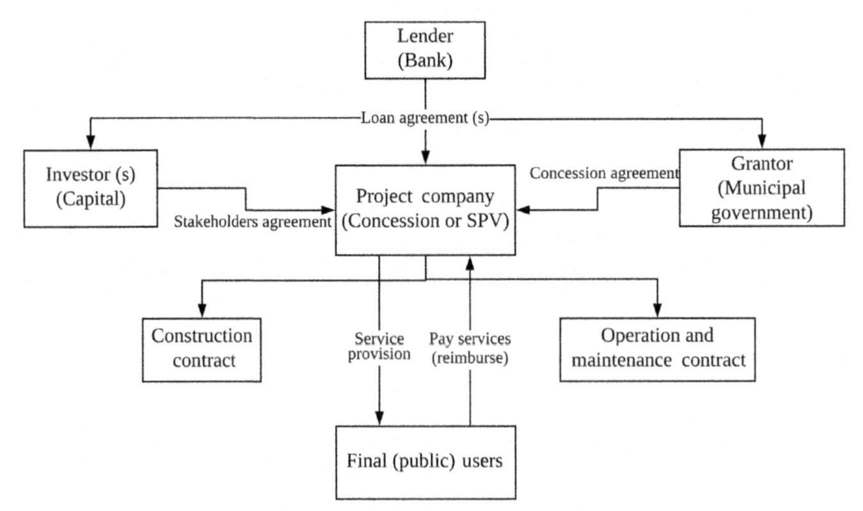

Figure 2.7 Typical project finance structure.

is building a metro line and has decided to adopt a project-financing scheme. The project company is first created and establishes a concession agreement with the grantor (the municipal government of the city). The project company then signs various stakeholder agreements with the government and private entities that are investing in the metro system. The project company then signs the loan agreement with the lender (usually a bank). Lastly, through tendering and bidding, construction and operation agreements are signed between the project's company and contractors and operators.

Project financing allows shareholders to keep project liabilities "off-balance," thus saving the debt capacity or fiscal space of shareholders for other investments. Meanwhile, this mechanism is also "nonrecourse" to shareholders, meaning the lender cannot take the assets and revenue of shareholders if the project's company fails to pay back loans.

2.6 RELIABILITY OF FORECASTING

Forecasting for underground structures is more complicated than for aboveground projects. For subsurface projects, according to Muir Wood (2000), "the inherent and highly site-specific nature of the ground adds one more uncertainty to the problems of forecasting into a future scene of unknowable political and economic circumstances."

There are numerous reasons a project's early estimates can change. The Channel tunnel, for instance, had significant changes in its project requirements, leading to an 80% increase in its cost estimation (Muir Wood, 2000). Although tedious, it is therefore essential to carry out a thorough risk analysis.

It is the responsibility of the planning team to make reliable forecasts. This must not only be done on an engineering level but also on financial and political levels. The uncertainties that these factors bring must be thoroughly addressed to ensure project success. Not doing so may cause high costs (both financial and other) or even lead to incompletion of the project.

The following are two examples of forecasting failures in real engineering projects:

- During the construction of the Singapore MRT (mass rapid transit), ground conditions along the tunnel changed so drastically that the construction encountered rock/soil interfaces seven times in a short distance of 434 m, which led to delay of the construction process.

- The Melamchi water supply project in Nepal. The contractor of this project failed to give enough consideration to disturbances caused to locals, tedious license approval processes, and other cultural and political factors. This ultimately led to the project not being completed by the first contractor. Negligence of the influence of the project on local populations played a key role here in the somewhat failure of the project.

2.7 CASE STUDY: TRANSNATIONAL RAILWAY UNDER THE HIMALAYAS

The proposed trans-Himalayan railway tunnel connecting China and India is discussed in this section (Bai et al., 2013). Trade between China and India currently passes through maritime, air, and overland mountain road routes. The fourth alternative proposed is a railway tunnel that would pass under the Himalayas. The case study discussed here analyzes the necessity of such a project based on the predicted trade volume and proposes three alignment alternatives.

2.7.1 Background

China and India have seen considerable economic growth. The two nations are overtaking developed countries in world GDP rankings and will be the first and third largest economies in the world in 2030, respectively (Goldman Sachs, 2007). The great economic growth they both enjoy has led to soaring imports and exports. By 2020, it is estimated that the bilateral trade route between the two nations will be the world's largest (Gupta & Wang, 2009).

But a major geographical barrier to this is the Himalayan mountain range. It has prevented direct transportation of goods and people overland since the beginning of human history. Still to this day, trade between China, India, and Nepal relies heavily on the long-distance marine route, weather-sensitive air route, and tortuous mountain roads. The railway project proposed would ensure more reliable, faster mass transportation.

To improve cultural exchanges and enhance trade, the Chinese and Indian governments redeveloped the old Silk Road, Nathu La Pass, in 2006. But this was not a major success, given its winter closures and limited trade volume. The Sino-Nepal highway is the only accessible trade link between China and Nepal, though it is near saturation and prone to large-scale geological disasters such as landslides and flooding. Fig. 2.8 shows the existing land and sea transport routes between the three nations.

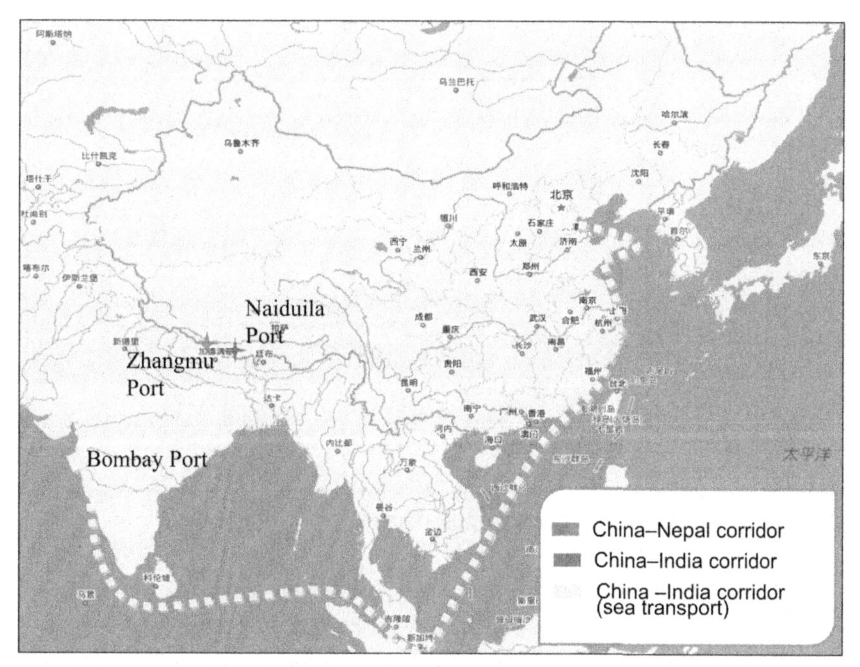

Figure 2.8 Transport routes between China, India, and Nepal.

For the benefit of trade, China and India need to be better connected. Nepal, lying between China and India, also needs development of its transportation infrastructures to help it gain economic independence.

It is imperative that the Himalayan railway project is safe, cost effective, and all-weather resistant. This will ensure the demand for future transport in the region is met and that trilateral cooperation between the three nations is strengthened.

2.7.2 Demand Projection

Passenger and freight transport demands are the main factors for decision makers and therefore need to be accurately predicted during the planning stage. In Fig. 2.9, it is assumed that the number of passengers from one country to another is linearly correlated to the GDP per capita. This assumption translates to saying that the wealthier the people, the more they will travel. The freight volume will not only depend on the bilateral trade volume, but also on the average unit price of freight. The methodology used to forecast the transport demand is illustrated in Fig. 2.9.

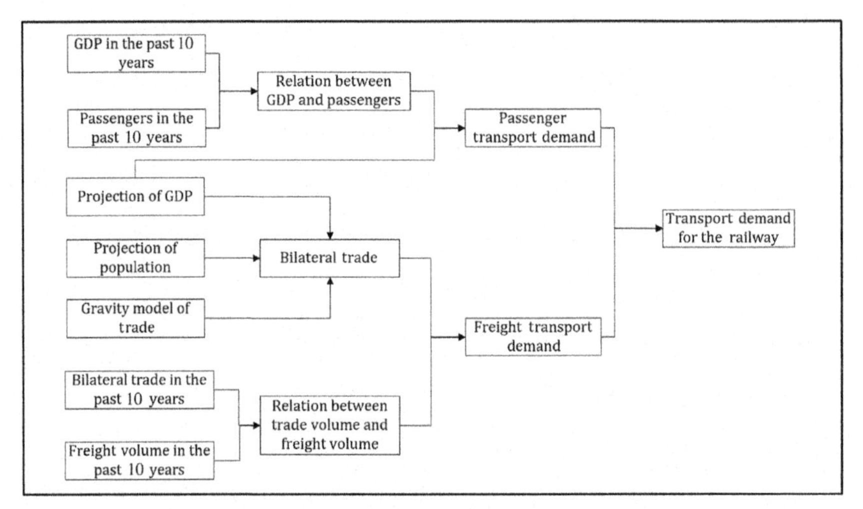

Figure 2.9 Procedure to forecast transport demand.

In this analysis, only trade between India and China are simulated. The Himalayan railway would in fact also be used by Nepal for passengers and freight transport to and from India and China. It would also serve western China by reducing transport times and cost. Lastly, the project would also generate a demand for subsequent railway projects in the region.

2.7.3 Alignment Selection

It is essential to become familiar with the geological features of the Himalayas region before proposing any alignment scheme. The Himalayas were formed by the collision of the Indian and Eurasian tectonic plates. The region is sheared and faulted due to the thrusting of the Indian plate. There are four major faults (Fig. 2.10), each of which is characterized by distinct geological features. Among them, the primary front is typified by folding, thrusting, and brittle fault zones. Formations in the main boundary thrust consist of quartzite and phyllite in brittle and ductile regimes due to intensive deformation. The central thrust is also a zone of metamorphic rocks. These rocks have extremely different properties, from homogenous to highly fractured with large quantities of water.

Given the geological and geomorphic conditions, three possible paths can be considered for the railway project. They are compared (Table 2.5) according to the design code for railway construction and using experience gained from similar projects in the past, such as those railways that

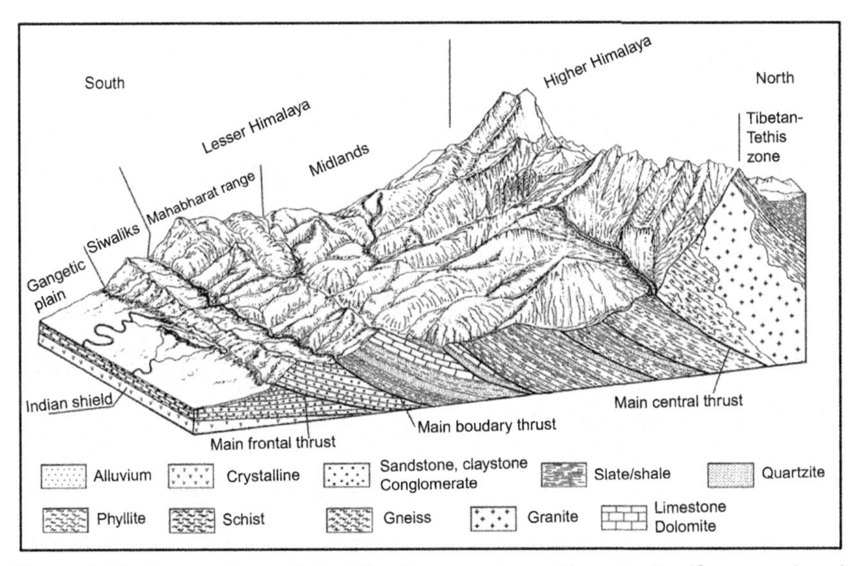

Figure 2.10 Cross-section of the Himalayan region with main landforms and rock units. *Adapted from Deoja, B., Dhital, M., & Wagner, A. (1991).* Risk engineering in the Hindu Kush Himalaya. *Nepal: International Centre for Integrated Mountain Development (ICMOD).*

Table 2.5 Comprehensive comparison of the three possible paths

Criteria	**Weighting of each criterion**	**West route**	**Middle route**	**East route**
Constructability	0.4	4	2	1
Economic benefit	0.2	3	2	4
Reliability	0.10	1	2	4
Operation efficiency	0.15	3	1	4
Environmental impact	0.15	2	1	4
Total score		3.05	1.7	2.8

cross the Alps and Andes. The normal gradient is set at 15% and the maximum at 30%, while the minimum curve radius is 300 m. The three paths share a common section from Shigatse to Xiamude and differ for the part between Xiamude and Kathmandu (Fig. 2.11):

- The west route would follow the Gyirong Valley after passing by the Peiku Lake. To meet the vertical gradient requirements, the route path would make loops to negotiate elevation difference. This would involve many tunnels and bridges, especially over the Nepalese section. The maximum overburden would be more than 2200 m.

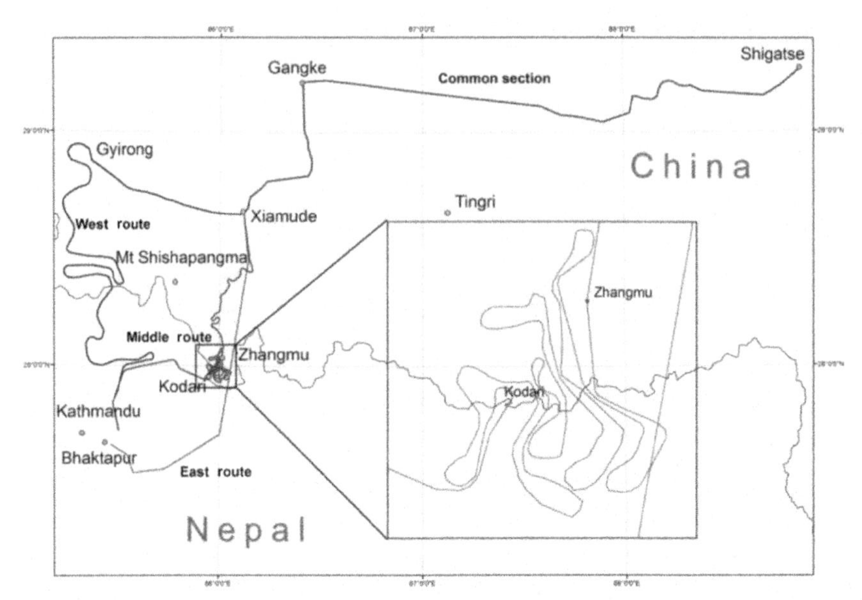

Figure 2.11 Three alternative alignments between Shigatse and Kathmandu.

- The middle route would follow the Zhangmu Valley after passing Nyalam and connect the important trading towns of Zhangmu and Kodari on the China–Nepal border. The proximity of the China–Nepal highway would provide easy access for construction and would benefit a greater number of people after the railway goes into operation. However, there is an important elevation drop (more than 1400 m) across a 30 km stretch, from Nyalam to Zhangmu. This would require many tight loops to ensure the maximum gradient is not exceeded (box in Fig. 2.11), while also complicating the construction process.
- The east route shares a common section with the middle route and would include a long base tunnel (130 km). Its average gradient is 30%. Although there would be more investments needed during the construction stage, the tunnel option would not only save resources and time but would also be more reliable and environmentally sustainable. Operation and maintenance costs would also be much lower compared to the other two routes.

2.7.4 Challenges and Countermeasures

Tunneling at high altitudes brings many challenges for engineers. Some of the most common are listed here with possible solutions to them. Technical difficulties are identified during the planning stage and must be seriously considered during the design stage of the project.

1. *Steep gradient*

   The Tibetan plateau is a plain with an average elevation of 4500 m and that of the Nepalese capital of Kathmandu is about 1700 m. Across the 130 km distance separating Xiamude and Kathmandu, the elevation drop is so drastic that it requires the construction of hairpin alignments or very long tunnels. The rugged topography further aggravates these problems.

2. *Bad geological conditions*

   As one of the youngest mountain ranges on earth, the Himalayas are still rising at a speed of 2 cm per year. Found from previous tunneling projects in Nepal and India, geological conditions are not favorable there. There are numerous faulted zones, sheared zones, and karst. Anisotropy and intercalation, which reduce rock strength, are two of the most important features of the Himalayan mountain range (Panthi, 2007). Swelling and squeezing of the rock is likely due to the high overburden of 2200 m. Due to high stresses within the rock, brittle areas are also prone to rock burst. This is a challenge for both conventional and mechanized tunneling in the design stage.

3. *Water ingress*

   Faulted zones and karst landforms in the Himalayan region are rich in underground water. This is a problem as high-pressure underground water can lead to fatal accidents in tunnels. Due to the rugged topography, it is impossible to thoroughly investigate the hydrogeology from the surface. The tunnel construction of this particular project therefore has a high level of risk. Predrilling techniques should thus be adopted to detect water in the tunneling face. Pregrouting could then also be used to seal potential water leakages.

4. *Permafrost*

   Permafrost and seasonal freezing soil is found almost everywhere on the Tibetan plateau. This may cause settlement and upheaval of the embankment. Construction should therefore be properly planned to minimize ground disturbances, and thermal insulation should be set up.

5. *Geothermal problem*

   The high overburden and active tectonic plates cause severe geothermal effects in the Himalayan region, leading to the existence of many hot springs. Tunneling through high-temperature regions is challenging for workers and can damage machinery through overheating.

Performant cooling systems therefore need to be designed to lower temperatures sufficiently.

2.8 QUESTIONS

2.1. List some of the reasons for cost overruns in tunneling projects.

2.2. Briefly introduce the different stages of planning for the use of subsurface spaces.

2.3. A tunneling project is a success if it is delivered on time, within budget, with the required quality and performance. Discuss this statement and list the reasons you agree or disagree.

2.4. How do you plan the use of a subsurface space? Which aspects need to be considered?

2.5. Briefly explain the difference between cost and price in the context of a tunneling project.

2.6. According to planning principles of underground structures, choose the right position of the tunnel according to its purpose and give the reasons.

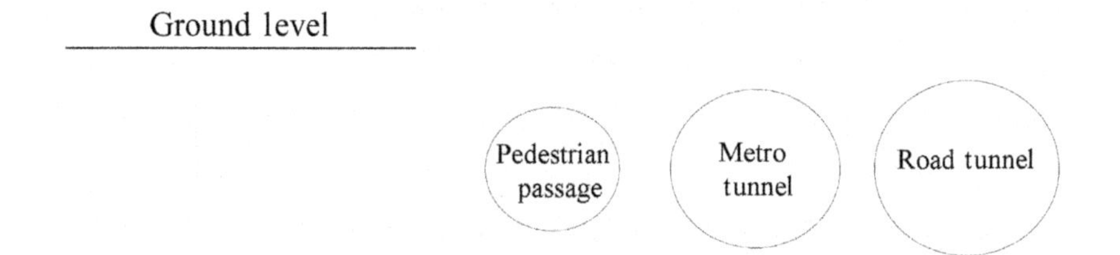

REFERENCES

Bai, Y., Shi, Z. M., Zheng, Y. L., Gu, F. F., Gu, L. L., & Dhital, M. R. (2013, April). *Trans-national railway under the Himalayas*. In *Paper presented at the 12th International Conference of Underground Construction*, Prague, Czech Republic.

Deoja, B., Dhital, M., & Wagner, A. (1991). *Risk engineering in the Hindu Kush Himalaya*. Nepal: International Centre for Integrated Mountain Development (ICMOD).

Fenalco. (2014). *Fondo Economico Nacional Colombiano, Bitácora Económica*. Retrieve from http://www.fenalco.com.co/bitacoraeconomica

Goldman Sachs. (2007). *BRICs and beyond*. Retrieved from http://www.goldmansachs.com/our-thinking/archive/archive-pdfs/brics-book/brics-full-book.pdf.

Gupta, A.K., & Wang, H.Y. (2009, September 1). *China and India: Greater economic integration*. Retrieved from https://www.chinabusinessreview.com/china-and-india-greater-economic-integration/.

INRIX. (2014). *Traffic congestion to cost the UK economy more than £300 billion over the next 16 years*. Retrieve from: http://inrix.com/press-releases/traffic-congestion-to-cost-the-uk-economy-more-than-300-billion-over-the-next-16-years/

ITA Committee on Underground Space. (2009). *White paper no.1, underground space Q&A*.

ITA Working Group no.4. (2000). Planning and mapping of underground space—an overview. *Tunnel and Underground Space Technologies*, *15*(3), 271–286.

ITA Working Group n°20. (2012, April). *ITA report no.011, report on underground solutions for urban problems*.

Labrit, G., Chatard, M., Walet, F., & Dupont, J. (2012). Major renovation of Croix-Rousse tunnel in Lyon from operation programme to design-build. *Tunnels et Espaces Souterrain*, *229*, 35–44.

Muir Wood, A. (2000). *Tunneling: Management by design*. London: E & FN SPON.

Panthi, K. K. (2007). Underground space for infrastructure development and engineering geological challenges in tunnelling in the Himalayas. *Hydro Nepal: Journal of Water, Energy and Environment*, *1*(1), 43–49.

PPPIRC. (2016). *Project finance—key concepts*. Retrieved from https://ppp.worldbank.org/public-private-partnership/financing/project-finance-concepts

Ronka, K., Ritola, J., & Rauhala, K. (1998). Underground space in land-use planning. *Tunnelling and Underground Space Technology*, *13*(1), 39–49.

Tibidibtibo. (2013). *Tunnel de la Croix-Rousse—Lyon* [diagram]. Retrieved from https://commons.wikimedia.org/wiki/File:Tunnel_de_la_croix_rousse.svg

Tunnels and Tunnelling. (2005). Retrieved from https://tunnelsonline.info

United Nations. (2014). *Report of the world urbanization prospect*. Retrieve from https://esa.un.org/unpd/wup/publications/files/wup2014-highlights.pdf

Vähäaho, I. (2014). Underground space planning in Helsinki. *Journal of Rock Mechanics and Geotechnical Engineering*, *6*(5), 387–398.

FURTHER READING

Transport for London. (2015). *Traffic note 4: Total vehicle delay for London 2014–15—Road space management outcomes, insight and analysis*. Retrieved from http://content.tfl.gov.uk/total-vehicle-delay-for-london-2014-15.pdf

CHAPTER 3

Design of Underground Structures

Contents

Design is a core task for civil engineers. The design of underground structures is different from those aboveground due to the additional geological conditions that must be taken into account. Adequate experience and theoretical analysis is therefore required for such projects (Fig. 3.1).

Underground Engineering
DOI: https://doi.org/10.1016/B978-0-12-812702-5.00003-7

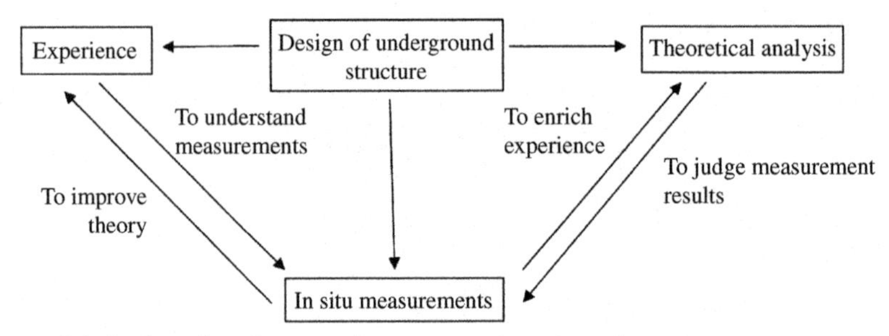

Figure 3.1 Design of underground structures: three key elements.

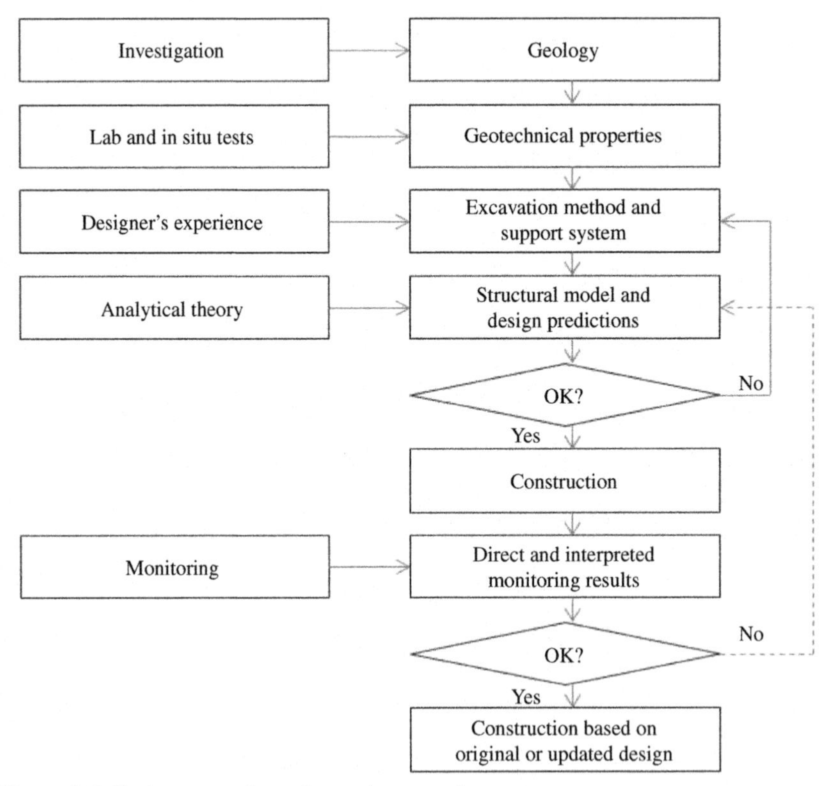

Figure 3.2 Design procedures for underground structures.

In this chapter, a basic introduction to underground engineering design is presented, with a focus on tunnels. First, classical design models and theories are discussed, as they are the core parts of the design process. Then, other stages in design are introduced including the site investigation prior to the modeling and calculation, as well as the instrumentation and back analysis during construction (Fig. 3.2).

As discussed in Chapter 1, History of Subsurface Development, the first man-made underground structures date back millennia. But it is not precisely known when tunnel construction evolved from heuristics (many of which that still remain have ingenious designs, such as the Karez in China). As one of the first known forms of underground engineering, its theories and techniques have been developed and improved. The following lists the most important underground engineering principles:

1. The ground may be a structural element of a tunnel.

Traditionally, the ground is regarded as an external load exerting pressure on a tunnel structure. Today this perception has slightly changed. Within acceptable deformation limits, some type of ground, especially rocks and stiff soils, are now considered a contributor to tunnel stability. This means the ground that surrounds a tunnel may be an integral part of its design.

2. It is important to select an appropriate support system.

With few exceptions, many underground openings risk collapse. Support systems (anchors, shotcrete lining, etc.) are needed to reinforce underground structures. An appropriate support system does not need to be cumbersome or heavy. In fact, a closed thin shell is more favorable given a phenomenon known as the ring effect. Flexible circular lining can deform under the uneven ground pressure, reducing the bending moments on the lining cross sections and thus reducing the risk of it being damaged. Other decisions related to the support system are the support type used and its installation time.

3. Instrumentation techniques are used to monitor displacements, loads, and stresses.

Due to uncertainties in the underground environment, instrumentation is needed at all times to monitor tunnel stability and to verify potentially unreliable geotechnical data. As with most engineering projects, underground design is a multidisciplinary and multistage process. The fields of geology, geotechnics, environmental science, soil and rock mechanics, structure calculations, electrics and electronics, and aeraulics and hydraulics are involved (ITA, n.d.).

3.1 DESIGN MODELS FOR UNDERGROUND STRUCTURES

Before introducing design methods and to better understand what needs to be designed, the characteristics of lining behavior need to be addressed.

- Before the lining gets in direct contact with the surrounding ground, any process such as ground pretreatment, excavation, and/or ground stabilization will affect the initial state of stress in the ground.
- The loads acting on a tunnel lining are not well defined, due to its dependency on the properties of the surrounding ground. The design of a tunnel lining is not a structural problem, but a ground–structure interaction problem, with emphasis on the ground.
- Tunnel lining behavior is said to be a four-dimensional (4D) problem due to several characteristics: (1) ground properties are time-dependent, leading to the common and critical phenomenon of standup time, without which most practical tunnel construction methods would be impossible; (2) transverse arching and longitudinal arching or cantilevering from the unexcavated face are possible ground conditions that can occur at the tunnel heading; and (3) the timing of lining installation is a critical variable.
- Not only serious structural problems, but also environmental problems encountered with actual lining behavior are related to the absence of support (voids left between the lining and the ground) rather than intensity and distribution of load. Therefore, construction goes hand-in-hand with design.
- The lining is a confined flexible ring. The deformation of the lining depends on the properties of the ground, therefore a proper criterion for judging lining behavior is the ductility to imposed deformations rather than the strength to resist bending stresses.

Numerous design techniques for underground structures have been developed, providing underground engineers with the tools to design safe and cost-effective support schemes. The support (the most common being reinforced concrete lining) is the tunnel's structure and should therefore resist external loads. Based on different perspectives and levels of complexity, most designs can be categorized into one of the following four types:

1. Empirical model: solely indications from past experience are used.
2. Load–structure model: actions from the ground are viewed as acting loads on underground structures and resistance from the peripheral ground layer (resistance of adjacent ground) is considered.
3. Ground–structure model: structures and ground are considered together as an interacting system and numerical calculations are often involved.

4. Convergence and confined model: characteristic curves of ground and support are used. This is effective in the design of underground openings in rock formations.

3.1.1 Empirical Models

Having already been used hundreds of times, empirical methods are design methods that rely entirely on past experience, and are widely used in many engineering diciplines. For a long time, empirical methods were combined with engineering intuition, the latter evolving through repeated practice in engineering projects. However, it was difficult to implement such qualifiable and unquantifiable skills. Empirical methods emerged with a view to quantifying such skills and experience.

The empirical methods of underground engineering involve several indices. Though each underground structure is unique, ground conditions can be represented by a universally applicable set of indices (such as rock mass raiting, or RMR, and Q-systems). Different sites that have similar indices usually require similar support schemes. They are calculated from observable ground characteristics such as RQD (rock quality designation), joint spacing, etc.

3.1.1.1 Rock Mass Rating

Bieniawski developed the RMR system during 1972–73. As an empirical method, the RMR has been continually evolving. More recent international standards have made further changes, and it is now applied in ground-related engineering projects worldwide (Bienwaski, 1989). It takes six parameters into account to classify a rock mass, as follows:

1. rock strength (uniaxial compressive strength of the intact rock);
2. RQD;
3. spacing of discontinuities;
4. condition of dicontinuities;
5. groundwater condition; and
6. orientation of discontinuities (dip and strike).

Don U. Deere an independent international consultant on dams, tunnels and underground power stations from Florida and Don W. Deere a geotechnical engineer from Longmont, Colorado introduced the RQD in 1967. It is a measure of the rock fracturing degree. Considering a drill sample, the RQD is defined as the percentage of sound rock cores that have a length equal to or greater than 100 mm over the total drill length (Fig. 3.3).

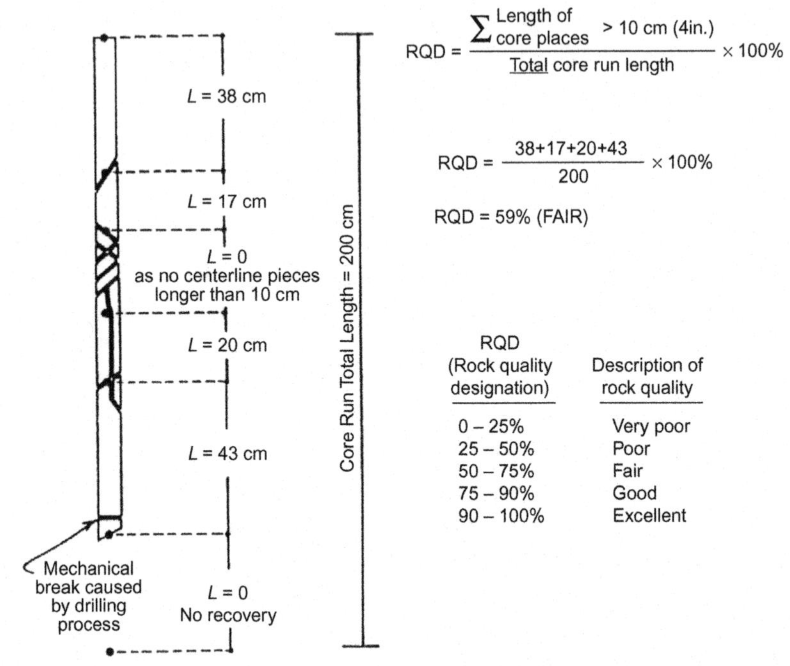

Figure 3.3 Procedure for measurement and calculation of RQD (Deere & Deere, 1988).

The rock mass refers to the rock material and rock discontinuities. The RQD does not account for the strength of the rock material or for the geometric properties of the joints. However, though this parameter has limitations in assessing the overall quality of rock mass, it is still widely used due to its simplicity. It is a geological parameter that can easily be observed and is key in determining the empirical RMR and the Q-value indices.

For the other parameters used in the RMR, joint spacing or spacing of discontinuities (sj) can be used to measure the average spacing between joints over a given length. Its inverse (the number of joints per meter of length) is defined as joint frequency (*ė*). It is obvious that the larger the joint spacing (or the lower the joint frequency), the better the rock continuity. Groundwater condition also significantly affects the design of the supporting system. One of the ways in which the impact can be studied is by measuring the inflow per 10 m tunneling length.

The RMR system takes the five basic parameters previously mentioned into consideration (Fig. 3.4). By evaluating them, individual

	Parameter		Ranges of values				
1	Strength of intact rock material	Point-load strength index (MPa)	>10	4 - 10	2 - 4	1 - 2	For this low range, uniaxial compressive test is preferred
		Uniaxial compressive strength (MPa)	>250	100 - 250	50 - 100	25 - 50	5 - 25 | 1 - 5 | <1
		Rating	15	12	7	4	2 | 1 | 0
2	Drill core quality RQD (%)		90 - 100	75 - 90	50 - 75	25 - 50	<25
		Rating	20	17	13	8	3
3	Spacing of discontinuities		>2m	0.6 - 2m	200 - 600mm	60 - 200mm	<60mm
		Rating	20	15	10	8	5
4	Condition of discontinuities		Very rough surfaces Not continuous No separation Unweathered wall rock	Slightly rough surfaces Separation <1mm Slightly weathered wall rock	Slightly rough surfaces Separation <1mm Highly weathered wall rock	Slickensided surfaces or Gouge <5mm thick or Separation 1 - 5mm Continuous	Soft gouge >5mm thick or Separation >5mm Continuous
		Rating	30	25	20	10	0
5	Groundwater	Inflow per 10m tunnel length (l/min)	None	<10	10 - 25	25 - 125	>125
		ratio (joint water pressure)/(major principal stress)	0	<0.1	0.1 - 0.2	0.2 - 0.5	>0.5
		General conditions	Completely dry	Damp	Wet	Dripping	Flowing
		Rating	15	10	7	4	0

Figure 3.4 Classification parameters and their ratings for RMR. *Bienawski, Z. T. (1989).* Engineering rock mass classifications: A complete manual for engineers and geologists in mining, civil, and petroleum engineering. *New York: Wiley, used by Hudson, J. A., & Harrison, J. P. (1997).* Engineering rock mechanics: An introduction to the principles *(1st ed.). Oxford, UK: Pergamon.*

parameter ratings are obtained. The RMR rating is given by the sum of all five-parameter ratings as shown below:

$$\begin{aligned}\text{RMR rating} &= \text{uniaxial compressive strength} + \text{RQD} + \text{joint spacing} \\ &\quad + \text{condition of joints} + \text{groundwater condition} \\ &\quad + \text{adjustments}\end{aligned} \tag{3.1}$$

After this, the joint orientation is taken into account using adjustments (Fig. 3.5).

As an empirical method the RMR can be used to estimate the standup time of an unsupported excavation (Table 3.1), which depends on both rock quality and tunnel size.

Fig. 3.6 depicts the relation between the RMR rating, the roof span, and the standup time.

B. GUIDELINES FOR CLASSIFICATION OF DISCONTINUITY CONDITIONS

Parameter	Ratings				
Discontinuity length (persistence)	<1m	1 - 3m	3 - 10m	10 - 20m	>20m
	6	4	2	1	0
Separation (aperture)	None	<0.1mm	0.1 - 1.0mm	1 - 5mm	>5mm
	6	5	4	1	0
Roughness	Very rough	Rough	Slightly rough	Smooth	Slickensided
	6	5	3	1	0
Infilling (gouge)	Hard filling			Soft filling	
	None	<5mm	>5mm	<5mm	>5mm
	6	4	2	2	0
Weathering	Unweathered	Slightly weathered	Moderately weathered	Highly weathered	Decomposed
	6	5	3	1	0

C. EFFECT OF DISCONTINUITY ORIENTATIONS IN TUNNELLING

Strike perpendicular to tunnel axis			
Drive with dip		Drive against dip	
Dip 45 - 90	Dip 20 - 45	Dip 45 - 90	Dip 20 - 45
Very favourable	Favourable	Fair	Unfavourable
Strike parallel to tunnel axis		Irrespective of strike	
Dip 20 - 45	Dip 45 - 90	Dip 0 - 20	
Fair	Very unfavourable	Fair	

D. RATING ADJUSTMENT FOR DISCONTINUITY ORIENTATIONS

Orientations of Discontinuities		Very Favourable	Favourable	Fair	Unfavourable	Very Unfavourable
Ratings	Tunnels & mines	0	-2	-5	-10	-12
	Foundations	0	-2	-7	-15	-25
	Slopes	0	-5	-25	-50	-60

Figure 3.5 Discontinuity orientation adjustments. *Bienawski, Z. T. (1989).* Engineering rock mass classifications: A complete manual for engineers and geologists in mining, civil, and petroleum engineering. *New York: Wiley, used by Hudson, J. A., & Harrison, J. P. (1997).* Engineering rock mechanics: An introduction to the principles *(1st ed.). Oxford, UK: Pergamon.*

3.1.1.2 Q-System

The Q-system was developed by Barton, Lien, and Lunde (1974) at the Norwegian Geotechnical Institute (NGI). The Q stands for quality. Just like the RMR, it has undergone modifications in recent decades before being defined as it is today. Although several geological parameters are required to determine the Q-system rating, the procedure is similar to that of the RMR rating. Six fundamental parameters are incorporated in the Q-system rating:

1. RQD;
2. joint set number;
3. joint set roughness;
4. joint alteration number;
5. joint water reduction factor; and
6. stress reduction factor (SRF).

Table 3.1 RMR ratings and meanings

	RMR ratings				
	100–81	**80–61**	**60–41**	**40–21**	**<20**
Rock mass class	A	B	C	D	E
Description	Very good rock	Good rock	Fair rock	Poor rock	Very poor rock
Average standup time	20 years for 15 m span	1 year for 10 m span	1 week for 5 m span	10 h for 2.5 m span	30 min for 1 m span
Rock mass cohesion (kPa)	>400	300–400	200–300	100–200	<100
Rock mass friction angle (degrees)	>45	35–45	25–35	15–25	<15

Source: Adapted from Bienawski, Z. T. (1989). *Engineering rock mass classifications: A complete manual for engineers and geologists in mining, civil, and petroleum engineering*. New York: Wiley, used by Hudson, J. A., & Harrison, J. P. (1997). *Engineering rock mechanics: An introduction to the principles* (1st ed.). Oxford, UK: Pergamon.

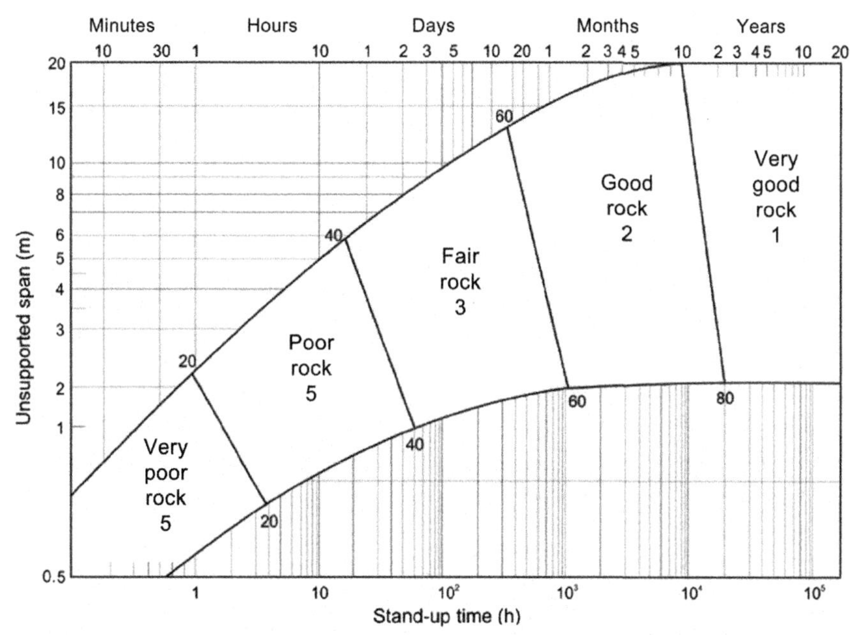

Figure 3.6 Relationship between the RMR rating and the standup time of unsupported excavation. *Abbas, S. M., & Konietzky, H. (2015). Rock mass classification systems. In H. Konietzky (Ed.),* Introduction to geomechanics. *Freiberg, Germany: TU Bergakademie Freiberg.*

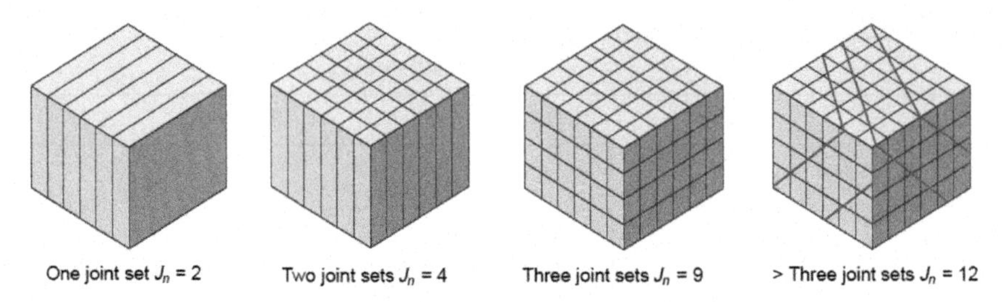

Note: the number of joint directions is not always the same as the number of joint sets

Figure 3.7 Different joint patterns and the equivalent joint set number (NGI, 2015).

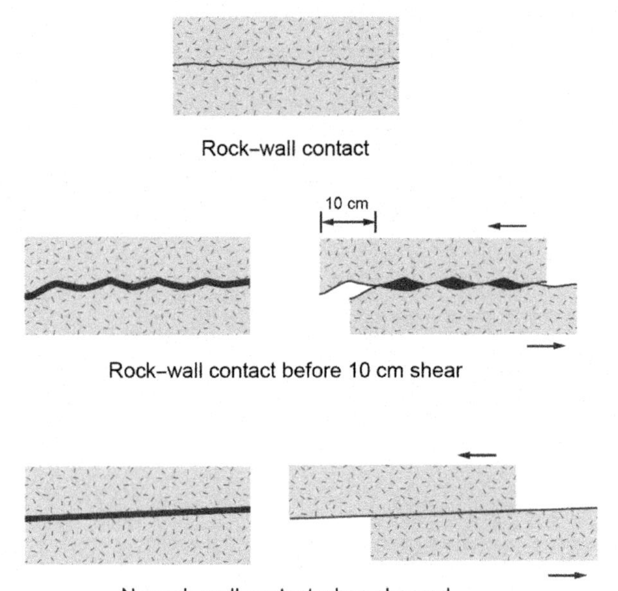

Figure 3.8 Joint roughness evaluation (NGI, 2015).

The RQD is the same as the one adopted in the RMR.

The number of joint sets (J_n) is a measurement of rock continuity (Fig. 3.7). Usually, one series of strongly developed parallel joints can be counted as one set, while occasional occurrences of breaks in cores are considered random joints.

Joint roughness number (J_r) is related to the roughness of the discontinuity surfaces (Fig. 3.8). The roughness is measured on the most unfavorable joint surface.

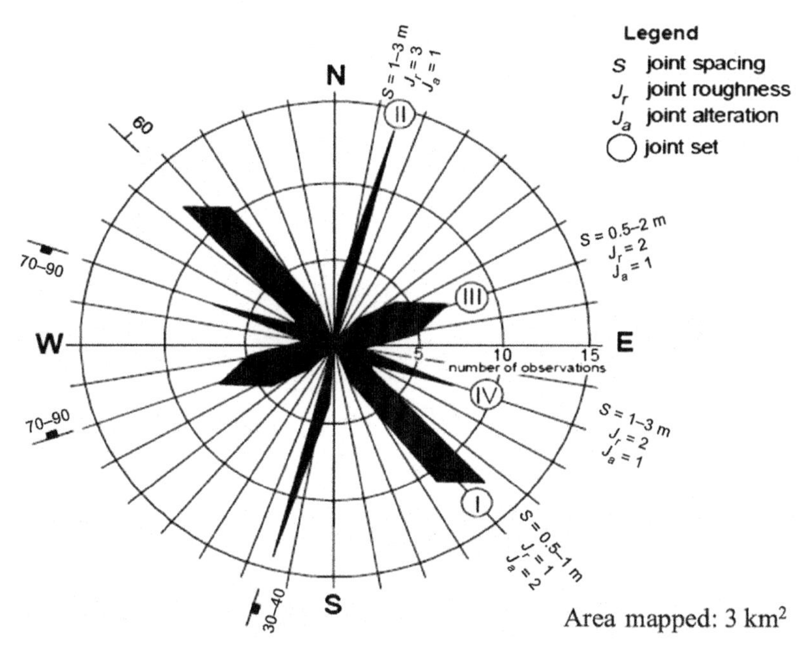

Figure 3.9 A joint rosette with four joint sets and corresponding J_r, J_a (Palmstrom & Broch, 2006).

The joint alteration number (J_a) represents the degree of alteration, weathering, or filling of discontinuity surfaces (Fig. 3.9). Alteration and clay filling have a negative effect on friction between joint surfaces. The higher the level of alteration and filling, the higher the value of J_a. As for the joint roughness number, the joint alteration number should be measured along the most unfavorable discontinuity.

The joint water reduction factor (J_w) serves to quantify the joint water condition. It is related to water pressure as well as the water inflow rate at discontinuities. This parameter is similar to the groundwater condition parameter used in RMR. The presence of water reduces normal stresses and softens and washes out clay in joints. Water therefore has an adverse effect on shear strength and rock stability.

The SRF reflects the presence of shear zones, stress concentrations, squeezing, and swelling rocks. These parameters all negatively contribute to tunnel stability. Fig. 3.10 shows shear zones in rock. Increasing potential faults in the rock, it is expected that their presence reduces the stability of excavation.

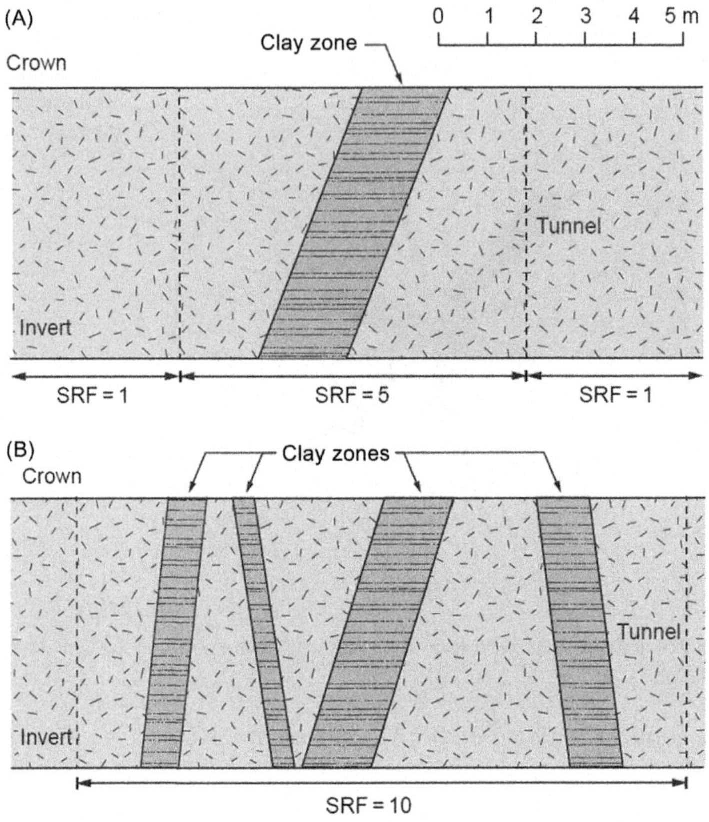

Figure 3.10 SRF in weakness zones: (A) with a single zone and (B) several nearby zones (NGI, 2015).

The Q-system rates (Fig. 3.11) and combines the six parameters mentioned earlier (Formula 3.2).

$$Q = \frac{\text{RQD}}{J_n}\frac{J_r}{J_a}\frac{J_w}{\text{SRF}} \tag{3.2}$$

These expressions help quantify tunnel stability and are measurements of:

1. the relative block size (RQD/J_n), the overall structure of the rock mass;
2. the interblock shear strength (J_r/J_a), the joint surface friction; and
3. active stresses (J_w/SRF).

1. ROCK QUALITY DESIGNATION (RQD)

		RQD
A.	Very poor	0—25
B.	Poor	25—50
C.	Fair	50—75
D.	Good	75—90
E.	Excellent	90—100

Note:
(i) Where RQD is reported or measured as ≤ 10 (including 0) a nominal value of 10 is used to evaluate Q in Eq. (1)
(ii) RQD intervals of 5, i. e. 100, 95, 90, etc. are sufficiently accurate

2. JOINT SET NUMBER (J_n)

		J_n
A.	Massive, no or few joints	0.5—1.0
B.	One joint set	2
C.	One joint set plus random	3
D.	Two joint sets	4
E.	Two joint sets plus random	6
F.	Three joint sets	9
G.	Three joint sets plus random	12
H.	Four or more joint sets, random, heavily jointed, "sugar cube", etc.	15
J.	Crushed rock, earthlike	20

Note:
(i) For intersections use $(3.0 \times J_n)$
(ii) For portals use $(2.0 \times J_n)$

3. JOINT ROUGHNESS NUMBER (J_r)

		J_r
	(a) *Rock wall contact* and (b) *Rock wall contact before 10 cms shear*	
A.	Discontinuous joints	4
B.	Rough or irregular, undulating	3
C.	Smooth, undulating	2
D.	Slickensided, undulating	1.5
E.	Rough or irregular, planar	1.5
F.	Smooth, planar	1.0
G.	Slickensided, planar	0.5
	(c) *No rock wall contact when sheared*	
H.	Zone containing clay minerals thick enough to prevent rock wall contact	1.0 (nominal)
J.	Sandy, gravelly or crushed zone thick enough to prevent rock wall contact	1.0 (nominal)

Note:
(i) Add 1.0 if the mean spacing of the relevant joint set is greater than 3 m
(ii) $J_r = 0.5$ can be used for planar slickensided joints having lineations, provided the lineations are favourably orientated

4. JOINT ALTERATION NUMBER (J_a)

		J_a	φ_r (approx.)
	(a) *Rock wall contact*		
A.	Tightly healed, hard, non-softening, impermeable filling i. e. quartz or epidote	0.75	(—)
B.	Unaltered joint walls, surface staining only	1.0	(25°—35°)
C.	Slightly altered joint walls. Non-softening mineral coatings, sandy particles, clay-free disintegrated rock etc.	2.0	(25°—30°)
D.	Silty-, or sandy-clay coatings, small clay-fraction (non-softening)	3.0	(20°—25°)
E.	Softening or low friction clay mineral coatings, i. e. kaolinite, mica. Also chlorite, talc, gypsum and graphite etc., and small quantities of swelling clays. (Discontinuous coatings, 1—2 mm or less in thickness)	4.0	(8°—16°)
	(b) *Rock wall contact before 10 cms shear*		
F.	Sandy particles, clay-free disintegrated rock etc.	4.0	(25°—30°)
G.	Strongly over-consolidated, non-softening clay mineral fillings (Continuous, < 5 mm in thickness)	6.0	(16°—24°)
H.	Medium or low over-consolidation, softening, clay mineral fillings. (Continuous, < 5 mm in thickness)	8.0	(12°—16°)
J.	Swelling clay fillings, i. e. montmorillonite (Continuous, < 5 mm in thickness). Value of J_a depends on percent of swelling clay-size particles, and access to water etc.	8.0—12.0	(6°—12°)
	(c) *No rock wall contact when sheared*		
K, L, M.	Zones or bands of disintegrated or crushed rock and clay (see G, H, J for description of clay condition)	6.0, 8.0 or 8.0—12.0	(6°—24°)
N.	Zones or bands of silty- or sandy clay, small clay fraction (non-softening)	5.0	
O, P, R.	Thick, continuous zones or bands of clay (see G, H, J for description of clay condition)	10.0, 13.0 or 13.0—20.0	(6°—24°)

Note:
(i) Values of $(\varphi)_r$ are intended as an approximate guide to the mineralogical properties of the alteration products, if present

5. JOINT WATER REDUCTION FACTOR (J_w)

		J_w	Approx. water pressure (kg/cm²)
A.	Dry excavations or minor inflow, i. e. < 5 l/min. locally	1.0	< 1
B.	Medium inflow or pressure occasional outwash of joint fillings	0.66	1.0—2.5
C.	Large inflow or high pressure in competent rock with unfilled joints	0.5	2.5—10.0
D.	Large inflow or high pressure, considerable outwash of joint fillings	0.33	2.5—10.0
E.	Exceptionally high inflow or water pressure at blasting, decaying with time	0.2—0.1	> 10.0
F.	Exceptionally high inflow or water pressure continuing without noticeable decay	0.1—0.05	> 10.0

Note:
(i) Factors C to F are crude estimates. Increase J_w if drainage measures are installed
(ii) Special problems caused by ice formation are not considered

6. STRESS REDUCTION FACTOR (SRF)

		SRF
	(a) *Weakness zones intersecting excavation, which may cause loosening of rock mass when tunnel is excavated*	
A.	Multiple occurrences of weakness zones containing clay or chemically disintegrated rock, very loose surrounding rock (any depth)	10.0
B.	Single weakness zones containing clay, or chemically disintegrated rock (depth of excavation ≤ 50 m)	5.0
C.	Single weakness zones containing clay, or chemically disintegrated rock (depth of excavation > 50 m)	2.5
D.	Multiple shear zones in competent rock (clay free), loose surrounding rock (any depth)	7.5
E.	Single shear zones in competent rock (clay free) (depth of excavation ≤ 50 m)	5.0
F.	Single shear zones in competent rock (clay free) (depth of excavation > 50 m)	2.5
G.	Loose open joints, heavily jointed or "sugar cube" etc. (any depth)	5.0

Note:
(i) Reduce these values of SRF by 25—50% if the relevant shear zones only influence but do not intersect the excavation

(b) *Competent rock, rock stress problems*

		σ_c/σ_1	σ_t/σ_1	SRF
H.	Low stress, near surface	> 200	> 13	2.5
J.	Medium stress	200—10	13—0.66	1.0
K.	High stress, very tight structure (Usually favourable to stability, may be unfavourable to wall stability)	10—5	0.66—0.33	0.5—2.0
L.	Mild rock burst (massive rock)	5—2.5	0.33—0.16	5—10
M.	Heavy rock burst (massive rock)	< 2.5	< 0.16	10—20

(c) *Squeezing rock;* plastic flow of *incompetent rock under the influence of high rock pressures*

		SRF
N.	Mild squeezing rock pressure	5—10
O.	Heavy squeezing rock pressure	10—20

(d) *Swelling rock; chemical swelling activity depending on presence of water*

		SRF
P.	Mild swelling rock pressure	5—10
R.	Heavy swelling rock pressure	10—15

(ii) For strongly anisotropic stress field (if measured): when $5 \leq \sigma_1/\sigma_3 \leq 10$, reduce σ_c and σ_t to $0.8\,\sigma_c$ and $0.8\,\sigma_t$; when $\sigma_1/\sigma_3 > 10$, reduce σ_c and σ_t to $0.6\,\sigma_c$ and $0.6\,\sigma_t$ where: σ_c = unconfined compression strength, σ_t = tensile strength (point load), σ_1 and σ_3 = major and minor principal stresses
(iii) Few case records available where depth of crown below surface is less than span width. Suggest SRF increase from 2.5 to 5 for such cases (see H)

Figure 3.11 Classification parameters and their original ratings for the *Q*-system (Barton et al., 1974).

But the RQD alone cannot represent a rock sample size. The joint set number (J_n) is used for this and thus used to overcome the limitations of the RQD.

The second ratio represents joint roughness and the contact condition between adjacent joint surfaces. Friction force is affected by both the friction coefficient and normal force between the surfaces in contact. If there is no contact between adjacent surfaces the friction force, which prevents the falling of rocks, will be close to zero. The parameters J_r and J_a are therefore interdependent and it is their ratio that is of interest.

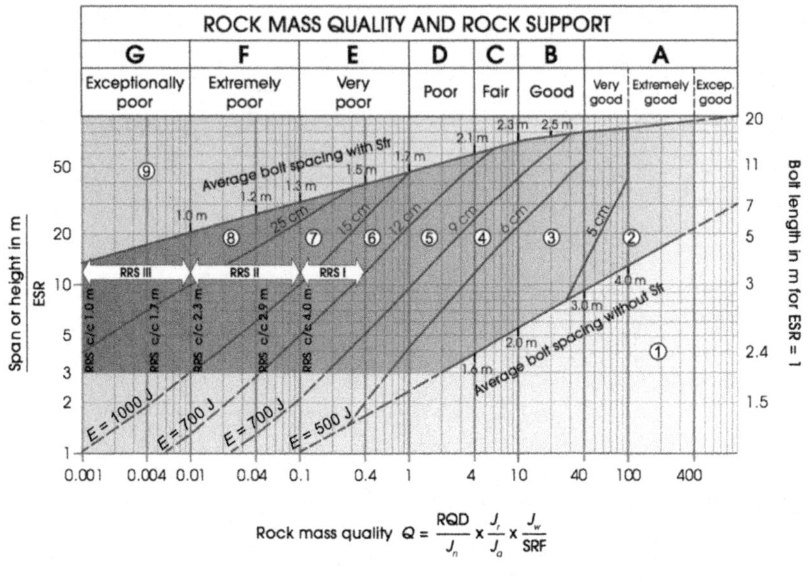

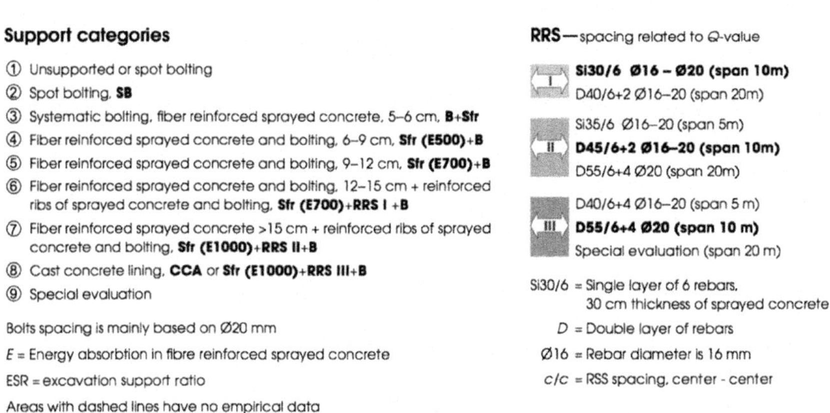

Support categories
① Unsupported or spot bolting
② Spot bolting, **SB**
③ Systematic bolting, fiber reinforced sprayed concrete, 5–6 cm, **B+Sfr**
④ Fiber reinforced sprayed concrete and bolting, 6–9 cm, **Sfr (E500)+B**
⑤ Fiber reinforced sprayed concrete and bolting, 9–12 cm, **Sfr (E700)+B**
⑥ Fiber reinforced sprayed concrete and bolting, 12–15 cm + reinforced ribs of sprayed concrete and bolting, **Sfr (E700)+RRS I +B**
⑦ Fiber reinforced sprayed concrete >15 cm + reinforced ribs of sprayed concrete and bolting, **Sfr (E1000)+RRS II+B**
⑧ Cast concrete lining, **CCA** or **Sfr (E1000)+RRS III+B**
⑨ Special evaluation
Bolts spacing is mainly based on Ø20 mm
E = Energy absorbtion in fibre reinforced sprayed concrete
ESR = excavation support ratio
Areas with dashed lines have no empirical data
RRS—spacing related to Q-value
I **Si30/6 Ø16 – Ø20 (span 10m)**
D40/6+2 Ø16–20 (span 20m)
II Si35/6 Ø16–20 (span 5m)
D45/6+2 Ø16–20 (span 10m)
D55/6+4 Ø20 (span 20m)
III D40/6+4 Ø16–20 (span 5 m)
D55/6+4 Ø20 (span 10 m)
Special evaluation (span 20 m)
Si30/6 = Single layer of 6 rebars, 30 cm thickness of sprayed concrete
D = Double layer of rebars
Ø16 = Rebar diameter is 16 mm
c/c = RSS spacing, center - center

Figure 3.12 The *Q* support chart (NGI, 2015).

The Q parameter is also a useful index to estimate the excavation stability and support requirements. It reflects the total support capability (temporary and permanent) of the roof structure (Fig. 3.12). Two additional parameters are necessary for its estimation: the equivalent dimension of the underground opening (D_e, Formula 3.3) and the excavation support ratio (ESR). The latter can be considered a safety factor that depends on the intended use of the excavation (Table 3.2).

Table 3.2 Various excavation support ratio categories

Excavation category	ESR
Temporary mine openings	3–5
Vertical shafts: circular section	2.5
Vertical shafts: rectangular/square section	2.0
Permanent mine openings, water tunnels for hydropower (excluding high-pressure penstocks), pilot tunnels, drifts, and headings for large excavations	1.6
Storage caverns, water-treatment plants, minor highway and railroad tunnels, surge chambers, access tunnels	1.3
Power stations, major highway or railroad tunnels, civil defense chambers, portals, intersections	1.0
Underground nuclear power stations, railroad stations, factories	0.8

Source: Adapted from Bienawski, Z. T. (1989). *Engineering rock mass classifications: A complete manual for engineers and geologists in mining, civil, and petroleum engineering*. New York: Wiley.

$$D_e = \frac{\text{Excavation span, diameter or height (m)}}{\text{Excavation support ratio (ESR)}} \tag{3.3}$$

Fig. 3.12 can also be extrapolated to determine tunnel wall support requirements if using the wall height and applying the following adjustments to Q:

- for $Q > 10$, $Q_{wall} = 5Q$;
- for $0.1 < Q < 10$, $Q_{wall} = 2.5Q$; and
- for $Q < 0.1$, $Q_{wall} = Q$.

In Fig. 3.12 it is possible to see that the Q-system values are more accurate estimations in the central part of the diagram. However, these empirical methods do have limitations.

3.1.1.3 Limitations, Links, and Extensions of Rock Mass Rating and Q

In real projects, tunnel engineers usually use both the RMR and Q-system parameters together. But it is important to be clear on their limitations:

1. RMR does not consider tunnel size and is only usable for tunnels of 3–10 m in diameter.
2. RMR does not differentiate roof and wall support.
3. For rock mass categories fair or above, the RMR and Q-system give similar support schemes, but the Q-system uses more shotcrete.
4. The Q-system is not suitable for applications in very poor rock so the RMR is adopted in these cases.

The empirical methods, and in particular the two indices developed here, are useful tools. The Q-system provides an estimation of tunnel support at the planning stage, namely for tunnels in hard and jointed rock masses without overstressing (Palmstrom & Broch, 2006).

Since the RMR and Q-system share parameters and work in a similar manner, some relations between the two indices have been proposed. Moreover, as rock mass properties are also related to intact rock properties and discontinuity properties, correlations have been established to estimate rock mass quality by using the RMR or Q-system.

Empirical methods are therefore useful to obtain an estimation of the rock quality and its meaning in terms of standup time and support requirements. But their ease of use is also a drawback, as they are based on experience gained from previous projects. This means they do not consider the uniqueness of new projects or complicated ground conditions.

Palmstrom and Broch (2006) in fact identified the use and misuse of rock mass classification systems, and concluded with a quote from Karl Terzaghi: *The geotechnical engineer should apply theory and experimentation but temper them by putting them into the context of the uncertainty of nature. Judgment enters through engineering geology.*

3.1.2 Load–Structure Model

Besides empirical models, other models exist that allow engineers to design tunnel supports in a more objective manner. The load–structure model is one of them. In this model, the tunnel lining is the structure to be designed, while the surrounding soil is treated as an external load exerted on the supporting structures.

The procedure of support design in the load–structure model is similar to the structure design of a building and contains the following steps:

1. determination of load conditions,
2. computation of internal forces, and
3. design of the structure cross sections.

3.1.2.1 Load Types

The loads that usually play an important part in the design of a tunnel lining include the vertical and lateral ground pressure, water pressure, self-weight of the lining, surcharge, and subgrade reaction.

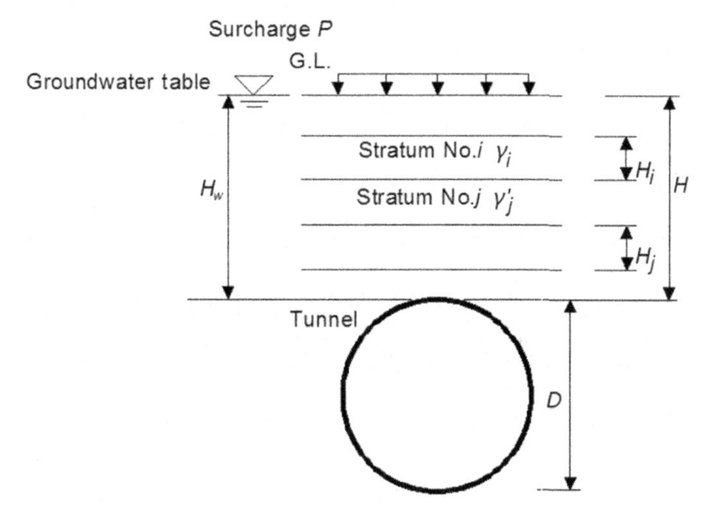

Figure 3.13 Section of tunnel and surrounding ground.

Vertical ground pressure (P_V) is induced by the self-weight of ground layers above the crown of the tunnel. For a shallow tunnel, the vertical earth pressure at the crown of the lining (P_V) equals the overburden pressure of the above layers, as indicated in Fig. 3.13 and Formula 3.4 (where γ_i and γ_j are the unit weight of soil stratum of the layer i and j located, respectively, above and below groundwater level, and P_o is the surcharge).

$$P_V = P_o + \sum \gamma_i H_i + \sum \gamma_j H_j \tag{3.4}$$

For a deep tunnel, the vertical ground pressure is reduced due to the arch effect or soil fraction. This reduced earth pressure (P_{e1}) can be calculated from Terzaghi's formula (see Formula 3.7).

Horizontal ground pressure (P_H) is the difference between earth pressure inside and outside the tunnel. This is similar to an earth-retaining wall where earth at the back of the retaining wall exerts a lateral force on it. To calculate the horizontal ground pressure of tunnels, the vertical ground pressure P_V is thus multiplied by the coefficient of lateral earth pressure K_0 at rest.

The combined effect of water pressure acting on the peripheral of the lining is the uplifting buoyancy. It should be compared with the total downward load. If the former is larger, the tunnel will float.

Self-weight of the lining is considered a *dead load* acting on the central line of the lining cross section. In other words, it is calculated as the

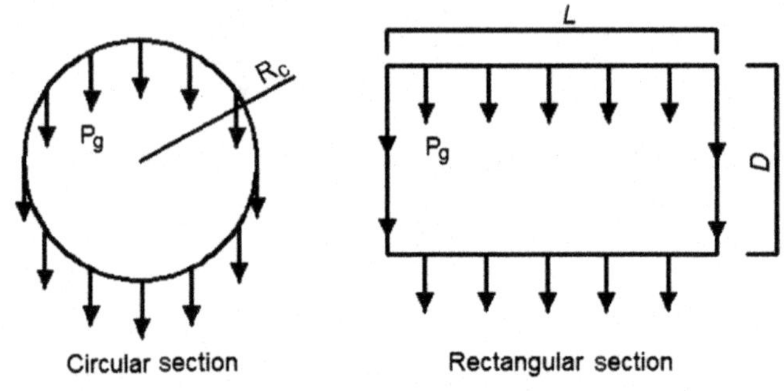

Figure 3.14 Self-weight of tunnel lining.

surface weight in kN/m^2. Formulas (3.5) and (3.6) show the calculation of the self-weight (P_g) of the lining as a function of the weight of the concrete linings per unit length (W), the radius of the central line of the lining cross section (R_c), the unit weight of the concrete (γ_c), and the thickness of the lining (t) (Fig. 3.14).

$$P_g = \frac{W}{2\pi R_c} \text{(circlular section)} \tag{3.5}$$

$$P_g = \gamma_c \times t \text{ (rectangular section)} \tag{3.6}$$

The surcharge is the result of traffic or structures above the tunnel. The Chinese design code identifies three types of surcharge: human load, road traffic load, and railway traffic load. Since they are *live loads*, both static and dynamic effects should be considered.

The subgrade reaction is the resistance from peripheral ground layers. It is a reaction force which depends on the ground displacement. It is assumed to be proportional to ground displacement with a proportionality factor that depends on the ground stiffness and lining dimensions (Fig. 3.15).

To understand this mechanism, the subgrade can be viewed as springs that are connected to the lining. When a circular tunnel section deforms under external loads and becomes an ellipse, the subgrade reaction occurs. However, it is important to note that the subgrade reaction and the ground pressure introduced at the beginning of this section are different load types. Ground pressure can be said to be active since it pushes a tunnel structure, whereas subgrade reaction is said to be passive as it is the result of a structure pushing the ground.

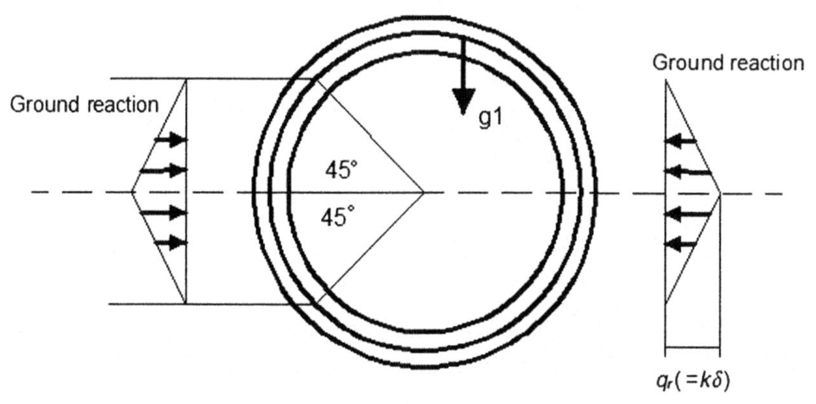

Figure 3.15 Subgrade reaction (soil reaction).

In addition to the abovementioned load types, other effects potentially also need to be considered, such as construction loads, seismic loads, the effect of an adjacent tunnel, etc.

3.1.2.2 Influence on Loads Caused by Tunnel Depth and Diameter

With increasing tunnel depths, the distribution and magnitude of ground pressure acting on the lining changes. As mentioned earlier, the calculation of ground pressure depends on the depth of the tunnel. This section focuses on ground pressure differences between shallow and deep tunnels.

Tunnels with a depth less than their diameter or constructed using the cut-and-cover method are considered shallow-buried tunnels. Such tunnels cause the soil above them to collapse when the tunnel itself collapses. When determining vertical ground pressure, the weight of all soil layers above the tunnel must therefore be considered. This ground pressure is estimated by the overburden pressure (Formula 3.4). The horizontal ground pressure is calculated through Rankine's lateral earth pressure theory. The subgrade reaction is at first assumed to take place at ± 45 degrees from the horizontal axis.

Tunnels with a depth of more than four times their diameter constructed using the drill-and-blast method or a TBM (tunnel boring machine) are called deep-buried tunnels. When this type of tunnel collapses, some of the soil above the tunnel crown will also (Fig. 3.16). Calculation methods include Terzaghi's formula (Formula 3.7) and

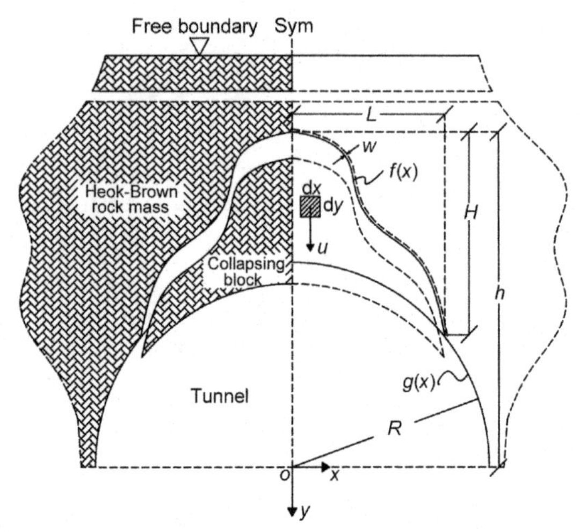

Figure 3.16 Collapsing pattern of deep tunnels (Zhang, Han, Fang, & Zhang, 2014).

Protodiaconov's formula. In addition to reduced vertical ground pressure, in the case of deep-buried tunnels this pressure is also similar in magnitude to the horizontal pressure.

$$
\begin{aligned}
P_{e1} &= \gamma h_o (\text{if tunnel is located above the groundwater table}) \\
P_{e1} &= \gamma' h_o (\text{if } h_o \leqq \text{Hw})
\end{aligned}
\tag{3.7}
$$

where P_{e1}, reduced earth pressure; h_o, reduced earth pressure divided by unit weight of soil.

For tunnel depths between one and four times the tunnel diameter, the arching effect varies with ground conditions, construction method, and overburden thickness. Past experiences of tunnels in similar conditions have been shown to be more valuable than methods involving theoretical calculation.

The load–structure model accepts that the action of ground is reflected on the underground structure only as loads (including active and passive soil pressure); the internal forces and deformation of are produced by the action of such load. The corresponding calculation method is called the load–structure method.

The calculation method given in *The Code for Design of Road Tunnels* (promulgated by the Ministry of Communications of the People's Republic of China on November 1, 2014) is used here as a reference.

3.1.2.3 Design Principle

The design principle of load–structure model is that the action of ground after tunnel excavation is mainly producing load on the lining structure, and the lining structure should bear the load including the ground pressure, etc., safely and reliably. The calculation first determines ground pressure using the ground classification criterion or practical formula, then calculates the lining's internal force as per the calculation method for structures on elastic foundation, and designs the structural cross section.

There are some assumptions in lining design:

1. The lining is perfectly flexible but capable of supporting appreciable ring stress in compression.
2. Shear stresses around the lining are negligible.

3.1.2.4 Analytical Solution for Soil Displacement Around Deep Excavation in Soft Clay

Terzaghi and Peck (1967) defined "deep excavation" as those excavations that exceed a depth of 6 or 7 m. According to following formula from limit analysis theory, it is obvious that vertical sides of excavation in soft clay must be supported.

$$H_{cr} = \frac{2 \cdot C}{\gamma} tg\left(\frac{\pi}{4} + \frac{\phi}{2}\right) \tag{3.8}$$

where H_{cr}, critical unsupported height of vertical sides (m); γ, bulk unit weight of soils, kN/m^5; C, soil cohesion, kN/m^2; φ, soil inner friction angel.

In the case of urban areas, where buildings are congested, diaphragm walling is used in most cases. Thus, the calculation of soil displacement around deep excavation supported by a propped diaphragm wall in soft clay formation will be introduced. The coefficients and factors that need to be strictly considered and controlled are as follows:

1. *Safety coefficient of base heave* (F_b)

   The following Terzaghi formula is still widely used in the condition of soft clay:

$$F_b = \frac{C_u \cdot N_c}{H(\gamma - C_u/0.7B) + q} (>1.5) \tag{3.9}$$

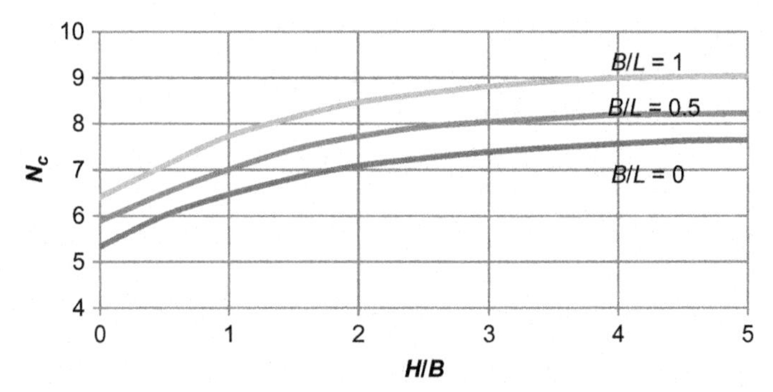

Figure 3.17 Coefficient N_c and depth–width ratio. *Modified from Mana and Clough (1981). Prediction of movements for braced cuts in clay.* Journal of Geotechnical Engineering, American Society of Civil Engineers, 107, *759–777.*

where C_u, shear strength, kPa; N_c, coefficient data from Fig. 3.17; H, depth of excavation, m; γ, bulk unit weight of soil, kN/m^3; B, width of trench, m; Q, surcharge, kPa.

The application scope of the above formula is $H/B < 1$. If $H/B > 1$, and the following Bjerrum and Eide formula can be used:

$$F_b = \frac{C_u \cdot N_c}{H \cdot \gamma + q} (> 1.5) \tag{3.10}$$

2. *Maximum lateral movement of diaphragm wall* (ΔH_{max})

   In 1981 Mana and Clough verified the relationship between the maximum lateral movement (ΔH_{max}) and the basal heave factor of safety (F_b) from field data by using the finite element method (FEM).
3. *Influence factors related to wall stiffness* ($E_w I_w / \gamma h^4$), *strut support stiffness* (S_k / γ_h), *hard layer depth* (D_w), *excavation width* (B), *and soil stiffness* (E/C_{uA})

   When the maximum lateral wall movement is calculated, several factors are not sufficiently considered such as wall stiffness ($E_w I_w / \gamma_h{}^4$), strut support stiffness (S_k / γ_h), hard layer depth (D_w), excavation width (B), and soil stiffness (H/C_u). Thus, Mana and Clough further introduced the values of these influence factors, which take into account such effects based on finite element studies. The values of these influence factors are shown in Figs. 3.18–3.20.

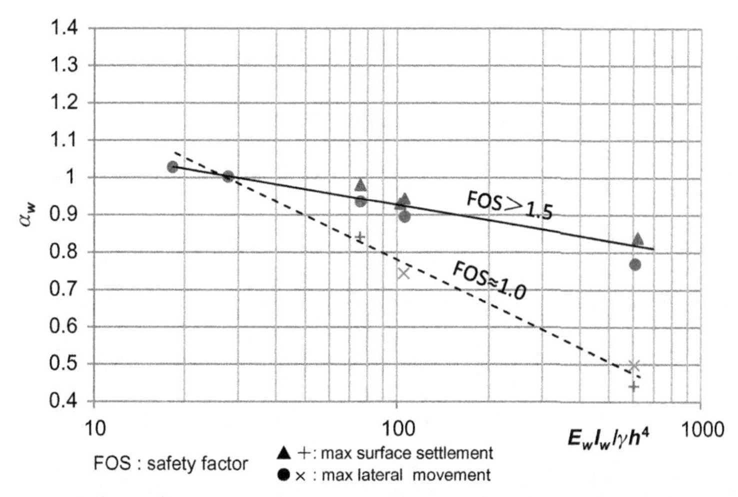

Figure 3.18 Relation between influence coefficient α_w and wall stiffness factor $E_wI_w/\gamma h^4$. *Modified from Mana and Clough (1981). Prediction of movements for braced cuts in clay.* Journal of Geotechnical Engineering, American Society of Civil Engineers, 107, *759–777.*

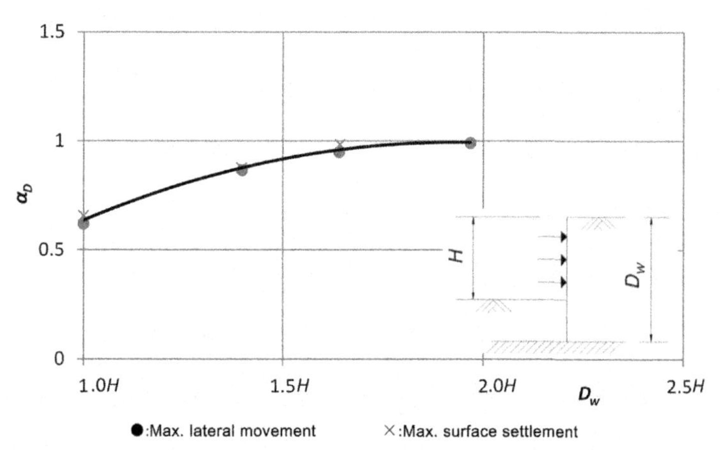

Figure 3.19 Relation between influence coefficient α_D and depth to firm D_w. *Modified from Mana and Clough (1981). Prediction of movements for braced cuts in clay.* Journal of Geotechnical Engineering, American Society of Civil Engineers, 107, *759–777.*

4. *Influence factors related to strut delay time* (a_{ti})

Based on experience in Shanghai, it is known that strut delay time will have a major influence on wall movement ($\Delta H_{\max}$). We use influence factor a_{ti} to reflect this influence. Table 3.3 shows the value

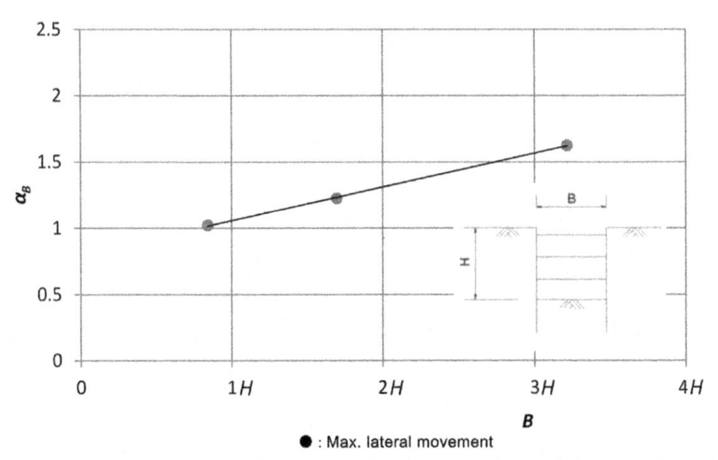

Figure 3.20 Relation between influence coefficient α_B and excavation width *B*. *Modified from Mana and Clough (1981). Prediction of movements for braced cuts in clay.* Journal of Geotechnical Engineering, American Society of Civil Engineers, 107, *759–777.*

Table 3.3 Strut delay time values on lateral wall movement

Layer number of strut	α_{ti} **(mm/day)**
Second layer of strut ($i = 2$)	2.3
Third layer of strut ($i = 3$)	2.4
Fourth layer of strut ($i = 4$)	3.4
Fifth layer of strut ($i = 5$)	4.1
Bottom slab ($i = B$)	1

of this influence factor (a_{ti}). The additional movement due to the delay of strut support time is

$$\sum \alpha_{ti} \cdot t_i$$

where t_i is the delay (in days) of different layer of strut.

5. *Influence factors of strut prestress* (α_p)

   Mana and Clough also studied the relation between strut prestress (α_p) and design stress using the FEM as shown in Fig. 3.21. In practice, strut prestress should be 80%, because stress loss will always happen to some degree. This conclusion is based on experience in Shanghai practice.

6. *Modified maximum lateral movement of diaphragm wall* (ΔH^*_{max})

   ΔH^*_{max} can be calculated using the following formula:

$$\Delta H^*_{max} = \Delta H^*_{max} \cdot \alpha_w \cdot \alpha_s \cdot \alpha_d \cdot \alpha_p \cdot \alpha_m \cdot \alpha_b + \sum \alpha_{ti} \cdot t_i \quad (3.11)$$

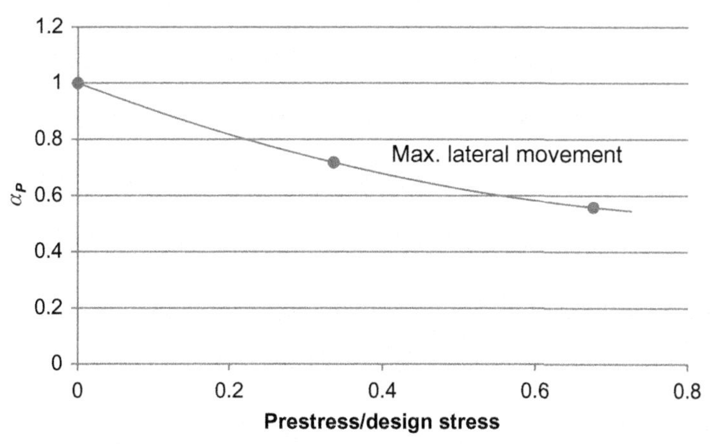

Figure 3.21 Effect on strut prestress on maximum lateral wall movement/design stress.

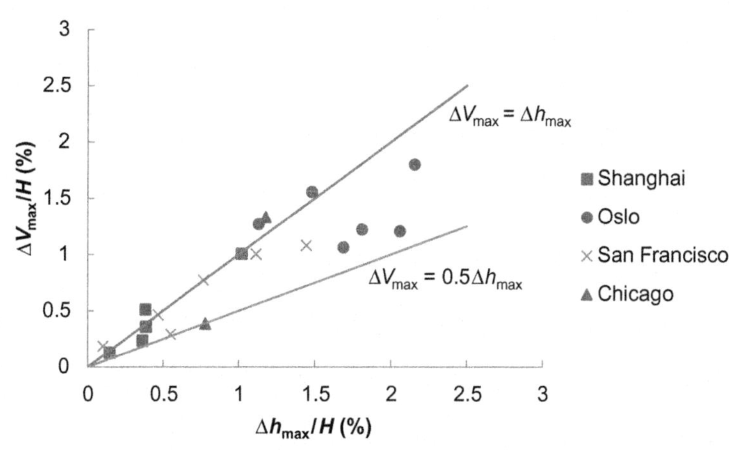

Figure 3.22 Case history data of the relationship between maximum ground settlements and maximum lateral wall movement. *Modified from Mana and Clough (1981). Prediction of movements for braced cuts in clay.* Journal of Geotechnical Engineering, American Society of Civil Engineers, 107, *759–777.*

7. *Maximum ground surface settlement* (ΔV^*_{max})

The relationship between ΔV^*_{max} and ΔH^*_{max} is shown in Fig. 3.22. As Fig. 3.22 shows, the following formula is reasonable:

$$\Delta V^*_{max} = \Delta H^*_{max} \tag{3.12}$$

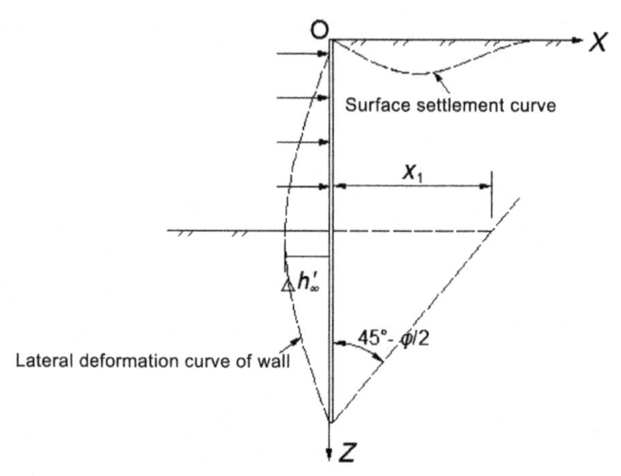

Figure 3.23 Diaphragm wall diagram.

8. *Ground surface settlement curve of cross section of diaphragm wall*

The ground surface settlement scope, X_0, can be calculated as follows:

$$X_0 = (H + D) \cdot tg\left(45^o - \frac{\phi}{2}\right) \qquad (3.13)$$

where H, depth of excavation, m; D, penetration of diaphragm wall, m; Φ, soil inner friction angle, degrees.

Based on Peck's assumption of the exponential function curve, the ground surface settlement curve of cross section of diaphragm wall is as follows:

$$\Delta V * (Z, X)|_{z=0} = a \cdot \left[1 - \exp\left(\frac{|X - 0.7H| + X_m}{X_0} - 1\right)\right] \qquad (3.14)$$

where X_0 can be found using Formula (3.13); Z, X, axes (Fig. 3.23); a, factor, based on Shanghai experience, as follows:

$$a = \Delta V^*_{\max} / \left[1 - \exp\left(\frac{0.7H}{X_0} - 1\right)\right], \qquad X_m = 0.7H \qquad (3.15)$$

9. *Longitudinal ground surface settlement curve along diaphragm wall*

Liou Jianhang studied the following formula, which can be used to calculate the scope (L) of longitudinal ground surface settlement based on Shanghai MRT station construction experience:

$$L = 2 - (H - h) * S + L_1 \qquad (3.16)$$

where H, excavation depth, m; h, max. depth of excavation where no settlement induced, usually 3–4 m; S, slope for one excavation section; L_1, length of one excavation section.

The shape of the longitudinal ground surface settlement curve is shown in Fig. 3.24. "a" is the length of circular arc with radius of "R," and "b" is the length of tangent line between two circular arc. "R" can be calculated based on the following formula:

$$R = \frac{L^2}{18 * \Delta V_{\max}^*} \tag{3.17}$$

Formula (3.17) can be used to predict the differential settlement of adjacent utilities, which are parallel to the diaphragm wall.

To better understand how the coefficients affect design a case study on a Shanghai stadium metro station, which was built by the cut-and-cover method, is described later.

Propped diaphragm walling was used as an excavation support system. The excavation has length (L) 232 m, depth (H) 14 m, and standard width (B) 22 m. The penetration depth of diaphragm wall (D) is 12 m.

The wall is supported by four strut layers. The elastic module of the wall (E_w) is 2.6×10^7 kPa. The thickness of the wall is 0.8 m. The stiffness of the wall ($E_w \cdot I_w$) is 1.109×10^6 kN-m^2/m. The outside diameter of the steel strut is 580 mm. The thickness of the cylindrical steel strut is 10 mm.

The horizontal spacing between two struts (d) is 3 m. The vertical spacing between two struts (h) is 3.5 m. The elastic module of the strut (E_g) is 2.1×10^8 kPa. The net cross-area is $A_g = 1.791 \times 10^{-2}$ m^2. The

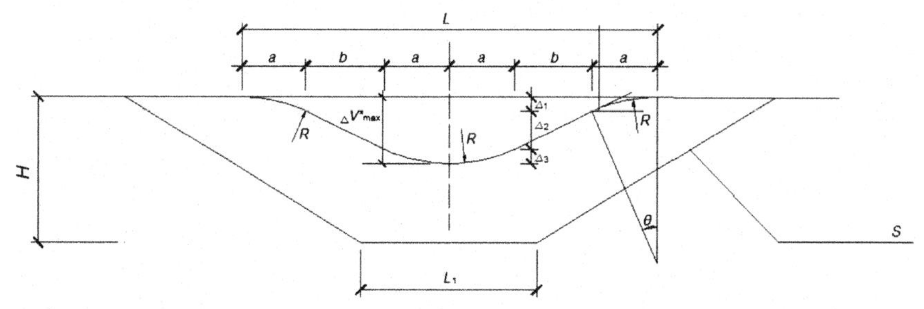

Figure 3.24 Longitudinal ground surface settlement.

stiffness of the strut (S_k) is $E_g \cdot A_g/(d \cdot h) = 3.582 \times 10^5$ kN/m^2. The pre-stress value of the strut is 70% of the designed stress.

The ground is gray muddy clay. The main parameters of the soil are $\gamma = 17.7$ kN/m^3, $C = 8.28$ kPa, $\Phi = 14.76$ degrees, $E = 4343$ kPa, and $C_u = \gamma \cdot H\ tg\Phi + C = 73.6$ KPa.

- Calculation procedure
  1. Base heave safety coefficient (F_b) calculation:

     Suppose: $q = 10$ kPa.

     From Formula (3.1),

$$F_b = \frac{C_u \cdot N_c}{H(\gamma - C_u/0.7B) + q}$$

$$\therefore F_b = \frac{73.6 * 7.6}{14 * [17.7 - 73.6/(0.7 * 22)] + 10} = 2.9 > 1.5$$

  2. Calculation $\Delta H^*{}_{\max}$ and $\Delta V^*{}_{\max}$:

     As $F_b = 2.9$, from Fig. 3.2 we can get:

$$\Delta H^*_{\max} = 0.43H(\%) = 0.43 \times 14 \times 0.01 = 0.06 \text{ m}$$

As $E_w \cdot I_w\ (\gamma \cdot h^4) = 418$, $S_k/(\gamma \cdot h) = 992$,

$$M = E/C_u = 4343/73.5 = 59,$$

From Fig. 3.17 to Fig. 3.22, we can calculate:

$\alpha_w = 0.84$, $\alpha_s = 0.76$, $\alpha_d = 1$, $\alpha_p = 0.53$, $\alpha_m = 1.7$, $\alpha_b = 1$ as, $t_i = 1(i = 2, 3, 4, 5, B)$, then:

$$\Delta H^*_{\max} = \Delta H^*_{\max} \cdot \alpha_w \cdot \alpha_s \cdot \alpha_d \cdot \alpha_p \cdot \alpha_m \cdot \alpha_b + \sum \alpha_{ti} \cdot t_i$$
$$= 0.06 \times 0.84 \times 0.76 \times 0.53 \times 1.7 \times 1 + 1 \times (2.3 + 2.4 + 3.4 + 4.1 + 1) \times 10^{-3}$$
$$= 0.0477 \text{ m(real result} = 0.052 \text{ m)}$$

The results are shown in Fig. 3.25.

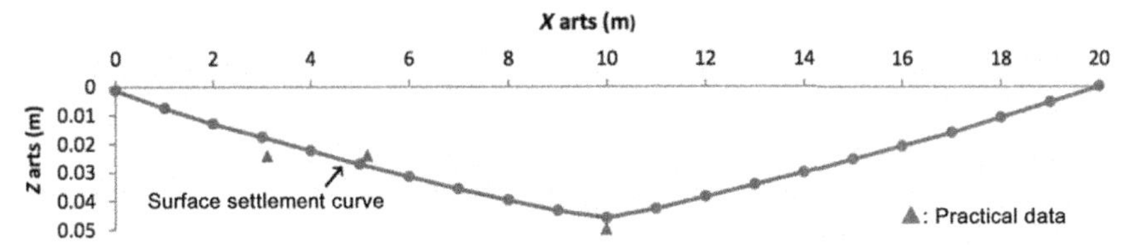

Figure 3.25 Practical and estimated ground surface settlement.

3.1.3 Ground–Structure Model

The ground–structure model includes both the ground and tunnel lining as the analytical objects. The method emerged from breakthroughs in construction techniques in the 20th century. Modern tunnel construction methods make it possible to install high-strength shotcrete lining rapidly and easily. This limits the deformation of the surrounding rock, thus preventing excessive broken-rock pressure that acts on the supporting structures. The surrounding rock is then no longer merely a burden but supports the tunnel structure itself, therefore becoming a supporting mechanism. The load–structure model is inapplicable in cases where shotcrete is used. This is because the tunnel structure and surrounding rock must be taken into consideration. This is the principle of the ground–structure model.

Two methods currently exist to deal with the ground–structure model. These are the analytical and numerical methods.

The first one involves the use of classical continuum mechanics to solve a problem. The second, using the FEM, allows designers to deal with more complex cases. It should be kept in mind that whichever method is adopted, the aim is to obtain the stress and deformation fields of the supporting structures and the surrounding rocks, based on which a tunnel structure design, optimizations, and construction processes can then be carried out.

The first step in the ground–structure model is to consider the interaction between the surrounding rock and tunnel structure. However, it is difficult to determine this analytically and is to this day a challenge for tunnelers. The ground–structure method therefore only serves as a complementary design method in most tunnel projects.

An example of a problem that can be solved using the analytical method of the ground–structure method is developed as follows.

A deeply covered tunnel is to be designed. The ground is considered an elastoplastic material and is subjected to a pressure P_0 at infinity. The contact pressure provided by the lining is P_i. A circular plastic zone has developed around the tunnel. Some known geotechnical parameters are as follows: the friction angle φ, the cohesion c, the elastic modulus E, and the Poisson's ratio ν.

The aim is to calculate the outer radius of the plastic zone and the radial displacement at the ground–structure interface.

Answer:

Suppose the radii of the lining's inner side, the lining-ground interface, and the plastic zone are R_1, R_0, and R_p respectively. An illustration of this scenario is shown in Fig. E.1.

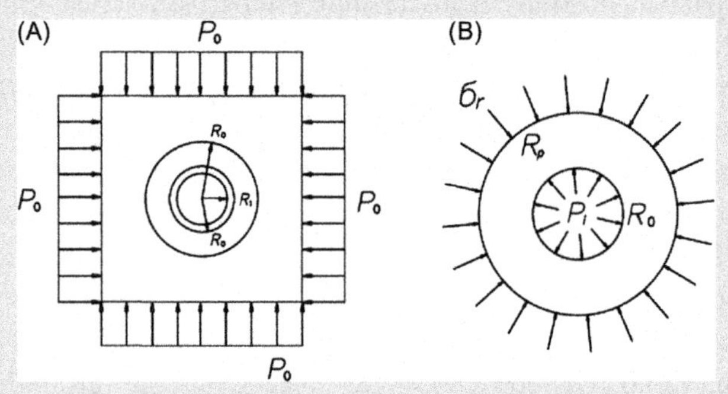

Figure E.1 An illustration of the tunnel cross section: (A) stress state of the surrounding rock and (B) stress state of the plastic zone.

1. Stress state of the plastic zone

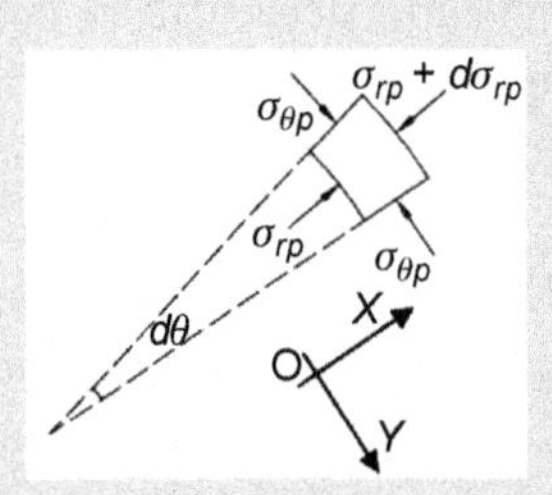

Figure E.2 Infinitesimal body at the plastic zone.

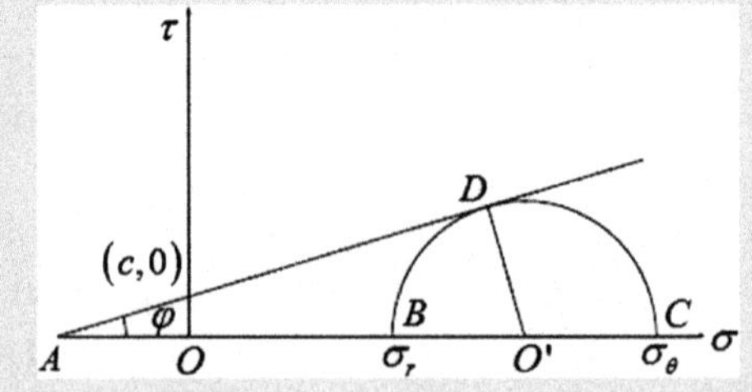

Figure E.3 Mohr–Coulomb diagram of the ground material.

(Continued)

(Continued)

Taking out an infinitesimal body in the plastic zone (Fig. E.2), the equilibrium condition requires

$\sum F_x = 0$, which expands into

$$\sigma_{rp} \cdot ds_{in} + 2\sigma_{\theta p} \cdot dr \cdot \sin\frac{d\theta}{2} = (\sigma_{rp} + d\sigma_{rp}) \cdot ds_{out} \tag{E.1}$$

where $ds_{in} = rd\theta$, $ds_{out} = (r + dr)d\theta$.

After manipulation, the equilibrium condition in Eq. (E.1) can be expressed as

$$\sigma_{\theta p} = r \cdot \frac{d\sigma_{rp}}{dr} + \sigma_{rp} \tag{E.2}$$

For the plastic zone, its stress circle on the Mohr–Coulomb diagram is tangent to the strength envelope line of the rock material (Fig. E.3). By solving the geometric relationship, the following is obtained:

$$\sigma_{\theta p} = \frac{1 + \sin\varphi}{1 - \sin\varphi}\sigma_{rp} + \frac{2c \cdot \cos\varphi}{1 - \sin\varphi} \tag{E.3}$$

Substituting Eq. (E.2) into Eq. (E.3) leads to a first-order linear differential equation:

$$\frac{d\sigma_{rp}}{dr} - \frac{2\sin\varphi}{1 - \sin\varphi} \cdot \frac{1}{r} \cdot \sigma_{rp} = \frac{2c \cdot \cos\varphi}{1 - \sin\varphi} \cdot \frac{1}{r} \tag{E.4}$$

The general solution is

$$\sigma_{rp} = -c \cdot \cot\varphi + C \cdot r^{\frac{2\sin\varphi}{1-\sin\varphi}} \tag{E.5}$$

To determine the constant C, the boundary condition at the support–ground interface is introduced:

$$\sigma_{rp}\Big|_{r=R_0} = P_i = -c \cdot \cot\varphi + C \cdot R_0^{\frac{2\sin\varphi}{1-\sin\varphi}} \tag{E.6}$$

So

$$C = \frac{P_i + c \cdot \cot\varphi}{R_0^{\frac{2\sin\varphi}{1-\sin\varphi}}} \tag{E.7}$$

Thus,

$$\sigma_{rp} = c \cdot \cot\varphi\left[\left(\frac{r}{R_0}\right)^{\frac{2\sin\varphi}{1-\sin\varphi}} - 1\right] + P_i \cdot \left(\frac{r}{R_0}\right)^{\frac{2\sin\varphi}{1-\sin\varphi}} \tag{E.8}$$

(Continued)

(Continued)

$$\sigma_{\theta p} = c \cdot \cot\varphi \left[\left(\frac{1+\sin\varphi}{1-\sin\varphi} \right) \left(\frac{r}{R_0} \right)^{\frac{2\sin\varphi}{1-\sin\varphi}} - 1 \right] + P_i \cdot \left(\frac{r}{R_0} \right)^{\frac{2\sin\varphi}{1-\sin\varphi}} \cdot \left(\frac{1+\sin\varphi}{1-\sin\varphi} \right) \tag{E.9}$$

2. Stress state of the elastic zone

According to the theory of elasticity, the general solutions for stresses in the elastic zone are

$$\sigma_{re} = \frac{B}{r^2} + A$$
$$\sigma_{\theta e} = -\frac{B}{r^2} + A \tag{E.10}$$

The boundary conditions are $\sigma_{re}|_{r=R_p} = \sigma_{rp}|_{r=R_p}$ and $\sigma_{re}|_{r\to\infty} = P_0$. Substituting them into Eq. (E.10) yields

$$A = P_0$$
$$B = \left\{ c \cdot \cot\varphi \cdot \left[\left(\frac{R_p}{R_0} \right)^{\frac{2\sin\varphi}{1-\sin\varphi}} - 1 \right] + P_i \cdot \left(\frac{R_p}{R_0} \right)^{\frac{2\sin\varphi}{1-\sin\varphi}} - P_0 \right\} \cdot R_p^2 \tag{E.11}$$

Substituting A and B into Eq. (E.10) leads to the stress state at the elastic zone.

3. Outer radius of the plastic zone

Using the boundary condition $\sigma_{\theta e}|_{r=R_p} = \sigma_{\theta p}|_{r=R_p}$, it is possible to obtain

$$R_p = \left[\frac{(P_0 + c \cdot \cot\varphi)(1-\sin\varphi)}{P_i + c \cdot \cot\varphi} \right]^{\frac{1-\sin\varphi}{2\sin\varphi}} R_0 \tag{E.12}$$

4. Displacement of the tunnel wall

By using Hooke's law, the strain field of the elastic zone can be expressed as a function of the stress field:

$$\varepsilon_{\theta e} = \frac{1}{E'}(\sigma_{\theta e} - \nu' \sigma_{re}) \tag{E.13}$$

where $E' = \frac{E}{1-\nu^2}$ and $\nu' = \frac{\nu}{1-\nu}$. Substituting Eq. (E.10) into Eq. (E.13), the following is obtained:

(Continued)

(Continued)

$$\varepsilon_{\theta e} = \frac{1+\nu}{E}\left[-\frac{B}{r^2} + (1-2\nu)P_0\right] \tag{E.14}$$

According to the theory of elasticity, the displacement field $U(r,\theta)$ can be obtained from the strain field. Since $U_{\theta e} = 0$ in this axisymmetric system, it is possible to get

$$\varepsilon_{\theta e} = \frac{1}{r}\left(\frac{\partial U_{\theta e}}{\partial \theta} + U_{re}\right) = \frac{U_{re}}{r} \tag{E.15}$$

Substituting Eq. (E.14) into Eq. (E.15),

$$U_{re} = \frac{1+\nu}{E}\left[-\frac{B}{r} + (1-2\nu)P_0 r\right] \tag{E.16}$$

In particular, at the boundary between the elastic and plastic zone, the radial displacement U_p is

$$U_p = \frac{1+\nu}{E}\left\{\begin{array}{l}(2-2\nu)P_0 - c\cdot\cot\varphi\cdot\left[\left(\frac{R_p}{R_0}\right)^{\frac{2\sin\varphi}{1-\sin\varphi}} - 1\right] \\ -P_i\cdot\left(\frac{R_p}{R_0}\right)^{\frac{2\sin\varphi}{1-\sin\varphi}}\end{array}\right\}R_p \tag{E.17}$$

As there is no established general formula for the displacement of the plastic zone, the solution of the radial displacement at the inner circumference of the plastic zone U_0 depends on the displacement field of the elastic zone. Assume there is only a change in shape but little change in volume when the ground material develops plastic deformation and that U_0 and U_p are sufficiently small. This yields

$$2\pi R_0\cdot U_0 = 2\pi R_p\cdot U_p \tag{E.18}$$

Substituting Eq. (E.17) leads to

$$U_0 = \frac{1+\nu}{E}\left\{\begin{array}{l}(2-2\nu)P_0 - c\cdot\cot\varphi\cdot\left[\left(\frac{R_p}{R_0}\right)^{\frac{2\sin\varphi}{1-\sin\varphi}} - 1\right] \\ -P_i\cdot\left(\frac{R_p}{R_0}\right)^{\frac{2\sin\varphi}{1-\sin\varphi}}\end{array}\right\}\frac{R_p{}^2}{R_0} \tag{E.19}$$

Although the ground–structure method is not yet mature, it is generally used in research and analysis of important or large projects. As mentioned earlier, the particularity of underground structures makes it necessary to use experience in the design of tunnel-supporting mechanisms.

Numerical models can supplement analytical modeling. As described, the nonelastic nature of rock, together with the discontinuities, heterogeneousness, anisotropic, and three-dimensional (3D) nature of rocks or ground material, hampers complete mechanical behavior analysis of such ground conditions. Therefore, numerical models are used to simulate detailed rock fracturing sequence in small and large scales.

3.1.3.1 Numerical Methods

Numerical methods have been adopted to design, analyze, and further estimate the deformations and behavior of underground structures. When an engineer is in charge of a new project, the more details he can obtain from its condition the more precise the calculation will be. Today, thanks to the growth of computers, underground engineers are able to adopt powerful numerical methods to analyze not only complex behaviors for both soil and rock medium but also make further estimations on the influences from different construction stages such as initial stress, ground behavior, excavation sequences, and support installation.

Stress deformation of different elements for different behavior of materials such as elastic, elastoplastic, viscoplastic, and strain softening can be incorporated into the analysis.

Almost every tunneling project requires numerical analysis to predict ground movement and to analyze the behavior of the surrounding ground in response to tunneling works. Among the numerical methods or numerical codes developed in recent decades and commonly used for underground excavation and tunneling are the FEM, finite difference method (FDM), and boundary element method (BEM), all based on the continuum approach. Table 3.4 summarizes the difference between FEM and FDM.

These methods facilitate modeling of complex structures, taking into account both construction and ground conditions that cannot be analyzed using analytical methods.

The FEM is used to analyze problems with material and geometric nonlinearities, complex boundary conditions, and nonhomogeneities. PHASES, MIDAS/GTX, PLAXIS, and UNWEDGE are some of the commercial computer programs implementing FEM available today. A FEM analysis involves the following steps:

1. discretization of the region of interest;
2. selection of displacement model;
3. definition of strain–displacement and stress–strain relations;

Table 3.4 FEM and FDM differences

Features	Finite element methods	Finite difference methods
Field quantities (e.g., stress, displacement)	Vary throughout each element in a prescribed fashion using specific functions controlled by parameters	Every derivative in the set of governing equations is replaced directly by an algebraic expression at discrete points in space
Computation method	Often combine the element matrices into a large global stiffness matrix	Regenerates the finite difference equations at each step of the calculation
Method to solve algebraic equations	Explicit	Often implicit or matrix-oriented schemes
Procedure	Divides the geometry into small elements and calculates the stresses and strains in those elements before assembling them back using theory of superposition	Divides the problem into small time steps and predicts the stresses and strains of the next time step based on the present time step using finite difference formulation
Variables	Defined within elements	Undefined within elements
Best for	Static loads, linear, small-strain problems	Dynamic problems, nonlinear, large-strain, physical instability

4. derivation of element equations;
5. assembly and incorporation of boundary conditions;
6. solutions of primary unknowns;
7. computation of secondary unknowns; and
8. interpretation of results.

FDMs are numerical methods for solving differential equations by approximating them with difference equations, in which finite differences approximate the derivatives. It is the oldest and simplest technique and is considered a good method for dynamics and large deformations. It is briefly described as follows:

- It requires knowledge of initial condition and/or boundary condition.
- The derivatives in the governing equation are replaced by algebraic expressions in terms of field variables (stress or pressure, displacement,

velocity), which are described at discrete points in space as nodes, but are not defined between the nodes or by elements.

- No matrix operations are required.
- The explicit method is generally used in which the solution is achieved by time-stepping using small intervals of time generating grid values at each time step.

The BEM involves the discretization of the interior or exterior boundaries only and consists basically of definition and solution of a problem entirely in terms of surface values of traction and displacement. The BEM is classified as direct and indirect depending on the procedure used to construct relationships between the tractions and the displacements. Indirect formulation is used in rock engineering problems particularly with respect to underground structures (Sharma, 2009).

Coupled FEBEM: The FEM and the BEM are two well-established numerical methods used for the analysis of underground openings. The advantages of both methods are utilized by adopting FEBEM in which finite elements are coupled with boundary elements (Sharma, 2009).

Some of commercial software used in underground engineering is listed in Table 3.5.

FLAC and PLAXIS are most commonly used by advanced geotechnical consultants. The general modeling procedure for all these programs consists of five major steps:

1. configuration of tunnel cross section and excavation type;
2. specification of support system and ground treatment;
3. definition of excavation and construction stages;
4. configuration of mesh data with terrain and strata; and
5. preparation of reports and output data.

Table 3.5 Commercial software use in underground engineering

Software	Method	Usage
FLAC (fast Lagrangian analysis of continua)	FDM	General FDM
ABAQUS	FEM	General FEM with some geotechnical relations
ANSYS	FEM	Mechanical/structural
PLAXIS	FEM	Geotechnical
SIGMA/W	FEM	Geotechnical
SEEP/W (FEM)	FEM	Seepage analysis
MODFLOW (FEM)	FEM	Groundwater modeling

It is important to note that all these programs require users to have a sound understanding of the underlying numerical models and constitutive laws.

Explicit and implicit methods are approaches used in numerical analysis for obtaining numerical solutions of time-dependent ordinary and partial differential equations, as is required in computer simulations of physical processes such as groundwater flow and the wave equation. Explicit methods calculate the state of a system at a later time from the state of the system at the current time, while implicit methods find a solution by solving an equation involving both the current state of the system and the later one. Mathematically, if $Y(t)$ is the current system state and $Y(t+\Delta t)$ is the state at the later time (Δt is a small time step), then, for an explicit method, Eqs. (3.18) and (3.19) should be used to find $Y(t+\Delta t)$.

$$\text{Explicit method } Y(t+\Delta t) = F(Y(t)) \tag{3.18}$$

$$\text{Implicit method } G(Y(t), Y(t+\Delta t)) = 0 \tag{3.19}$$

It is clear that implicit methods require extra computation (solving the above equation), and they can be much harder to implement. Implicit methods are used because many problems arising in real life are stiff, for which the use of an explicit method requires impractically small time steps Δt to keep the error in the result bounded (see numerical stability). For such problems, to achieve given accuracy, it takes much less computational time to use an implicit method with larger time steps, even taking into account that one needs to solve an equation of the form (Eq. 3.19) at each time step. In conclusion, whether one should use an explicit or implicit method depends on the problem to be solved. Table 3.6 compares these two methods.

3.1.4 Convergence and Confined Models

The convergence and confined models are similar to that of the ground–structure model, in that they consider the interaction between the surrounding rock and supporting structures. Instead of tackling the problem through classical mechanics and numerical analysis, this model solves it by using the characteristic rock mass and support curves (Fig. 3.26). This is why the model is also known as the characteristic line method.

Table 3.6 Comparison of explicit and implicit solution methods (FLAC manual, Itasca)

Explicit	Implicit
Time step must be smaller than a critical value for stability	Time step can be arbitrarily large, with unconditionally stable schemes
Small amount of computational effort per time step	Large amount of computational effort per time step
No significant numerical damping introduced for dynamic solution	Numerical damping dependent on time step present with unconditionally stable schemes
No iterations necessary to follow nonlinear constitutive law	Iterative procedure necessary to follow nonlinear constitutive law
Provided that the time step criterion is always satisfied, nonlinear laws are always followed in a valid physical way	Always necessary to demonstrate that the abovementioned procedure is: (1) stable and (2) follows the physically correct path (for path-sensitive problems)
Matrices are never formed. Memory requirements are always at a minimum. No bandwidth limitations	Stiffness matrices must be stored. Ways must be found to overcome associated problems such as bandwidth. Memory requirements tend to be large
Since matrices are never formed, large displacements and strains are accommodated without additional computing effort	Additional computing effort needed to follow large displacements and strains

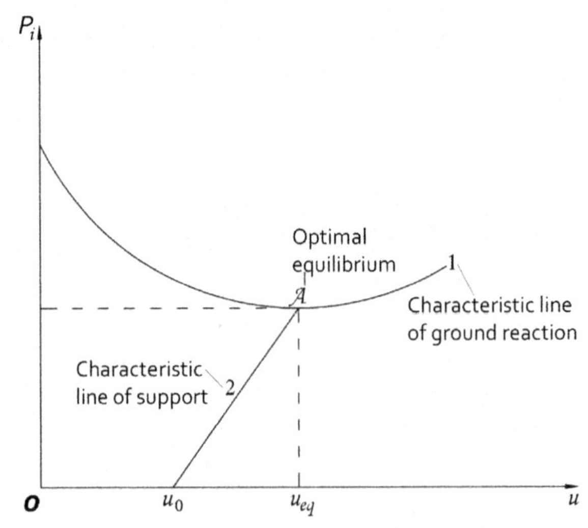

Figure 3.26 Characteristic line method in the convergence and confined model.

The method focuses on a two-dimensional (2D) plane strain problem of the ground–support interaction instead of a 3D one. When a stress is applied to the opening of the wall (known as confinement loss), it will decrease in magnitude and so the ground loses its confinement. This in turn leads to a displacement of the opening walls (Panet et al., 2001).

In Fig. 3.26 the horizontal axis denotes the radial displacement of the tunnel lining, and the vertical axis the contact pressure between the rock mass and the supporting structures. They are usually based on measurements at the tunnel crowns. For line 1, the y-coordinate corresponds to the constraint pressure exerted on the surrounding rocks by the support structures. Line 1 is for this reason also called the support demand curve of the surrounding rocks. For line 2, the y-coordinate corresponds to the constraint pressure provided by the supporting structures to stabilize the rock mass. Line 2 can therefore also be called the support supply curve of the tunnel supports.

When line 1 and line 2 intersect at point A, the deformation of the rock mass and the support are compatible and the contact pressure reaches the equilibrium. The y-coordinate of P_i is the true contact pressure between the supports and the rock mass, and the x-coordinate is the true tunnel wall displacement.

Since there is a time delay for support installation after tunnel excavation, the x-intercept of line 2 is not zero but u_0, which denotes the free tunnel wall deformation prior to the installation of the lining. By adjusting u_0, tunnel designers can control the intersection of the two characteristic lines at the lowest point of line 1, which results in the lowest contact pressure to stabilize the rock mass, thus giving the most economical tunnel structure design.

3.1.5 Tunnel Information Modeling

Before introducing tunnel information modeling (TIM), building information modeling (BIM) should be addressed, as TIM actually evolved from BIM. In current tunneling practice, finite element simulations have become an integral part of planning and design processes. These models are most often manually generated using 2D CAD drawings, which is a laborious and time-consuming process. Today, based on BIM approach, a 3D FE model is automatically generated through a set of compatible geometries of individual components, that is, the geology, alignment, lining, and the TBM. The BIM model also includes all relevant model parameters of the tunneling project that can then be incorporated into subsequent analysis during the tunnel drive. Set up and execution of FE analysis is performed automatically utilizing all required data from BIM.

3.1.5.1 Database Management

As with all modern large-scale construction projects, an enormous amount of data is collected and exchanged during a TBM drive. Some data is relevant for the creation of a FE model, and some is necessary for verification purposes. This data stems from a variety of sources and can be broadly divided into the following categories: site data, design data, machine data, monitoring data, and simulation data.

The following provides a short but nonexhaustive description of the many data sources encountered, how they are managed, and in what way they are relevant to an FE model.

- Site data: Site data is gathered from georeferenced borings that are conducted before the tunneling drive. A geological cross section is constructed using boring logs and, often, laboratory and field tests are performed in order to attain the mechanical parameters of the soil. Information on the water table along the planned tunnel route is gathered, and any additional data is gathered from local experience. These data are most often summarized in a geotechnical data report in which the data is stored in tabular (test and measurement protocols, mechanical soil properties), graphical (measurement charts, drilling profiles, geological cross sections), and text formats.
- Design data: The physical design data of the tunnel, such as tunnel alignment, minimum curve radius, tunnel diameter, and designed cross sections are often stored in the form of CAD drawings. Calculations and analysis of the tunnel lining and support systems as well as installation protocols and grouting pressure calculations are stored in tabular or in text format. These data, along with site data, form the basis for the creation of a typical FE model. Logistical factors, such as environmental aspects, legal aspects, and economic factors, also affect the design and construction and are typically stored in a document format.
- Machine data: Due to its many interacting components, a TBM produces a large amount of data. This data includes the thrust forces, the torque and rotational speed of the cutting wheel, the advance speed of the machine, the face support pressure, the grouting pressure, and the monitoring of any system alarms that may sound during the drive.

The data is recorded as instantaneous data measured in 10 seconds intervals, and as average values over one advance ring. These values lead to about 1.7–3.5 million accumulated data points per day (Miranda et al., 2011), typically stored in tabular format. Additionally, protocol documents are logged by machine operators and stored in document format. The collection and analysis of these data are important for the validation of more complex FE models.

- Monitoring data: Monitoring schemes usually involve the measurement of displacements at certain predetermined points along the tunnel alignment at particular intervals. It can be generated using any number of measurement techniques and, due to large data storage capacities, these measurements are typically taken in very short intervals and provide several thousand data points per hour. These are stored in tabular format. The collection and analysis of these data are important in validating the predictions made by FE models.
- Simulation data: The results of numerical simulations provide a large number of data points that can be directly compared with monitoring data, and/or other calculations. This data is stored in separate file types based on the chosen simulation program. This data is accessed through a graphical NURBS-based model during the postprocessing of the results. These models are most often generated uniquely for each analysis, and the process of creating such a model is often quite time-consuming.

It is evident from the descriptions above that, not only is a large amount of data collected during a TBM drive, but that the data is highly heterogeneous and stored in a number of different formats. Current practice is to store this data locally with different parties such as with the contractor, the designer, or the owner, which results in confusing and complex data interaction chains. It is therefore clear that a central data management scheme is advantageous for large-scale mechanized tunneling construction projects.

3.1.5.2 Tunnel Information Model

BIM methods address the problems generated by decentralized data management, and use standardized exchange formats such as the Industry Foundation Classes (IFCs) to ensure that a coherent data exchange exists between all models and information sources within a project (Building Smart, 2015). BIM models organize data on geometrical and spatial levels and, by modifying IFCs, are able to easily augment a main model with project-specific elements. Such an element typically consists of a visual component that is linked to the main model geometry and an information component that is linked to the element geometry. Information is always accessed through a geometrical model and is intuitively organized. Additionally, BIM concepts are able to address the entire lifecycle of a building model, from planning to operation stages, which is critical for highly process-oriented projects, such as tunneling. Although BIM methods have most often been applied to buildings, they are currently also used for bridge and road projects and have also been applied to tunneling projects. An academic BIM model tailored to fit the needs of a tunneling project has been implemented using data taken from the Wehrhahn-line project in Düsseldorf, Germany.

The TIM includes tunneling-related geometrical models (tunnel, TBM, boreholes, ground and city models), property and city data, and measurements (machine data and settlement). Not only does the TIM provide a data management platform, but it also allows the user to visually interact with and analyze the data through animations or by sequentially time-stepping through processes.

Typically, the results of numerical simulations are not reflected in the construction stages of a TIM model, as often only the final structure is simulated. With respect to tunneling projects, such a methodology is often detrimental as settlements predicted by a simulation change due to deviations from the design during the construction stage. Furthermore, often only certain "problem areas" are subject to inspection through numerical analysis, so a simulation of the entire project domain is almost never performed. Current TIM methods satisfy these unique needs as they present dynamic simulation results.

3.2 SITE INVESTIGATION

One of the biggest differences that distinguish underground engineering from others of civil engineering is the enormous uncertainty associated with the underground construction environment. Underground space can hide many unpleasant surprises, such as karst caves (Fig. 3.27) or water-bearing faults. Besides the possibility of encountering adverse conditions, even basic soil and rock properties are difficult to determine. Without comprehensive knowledge of the underground conditions, it is impossible to properly proceed into the later stages of design and construction. Site

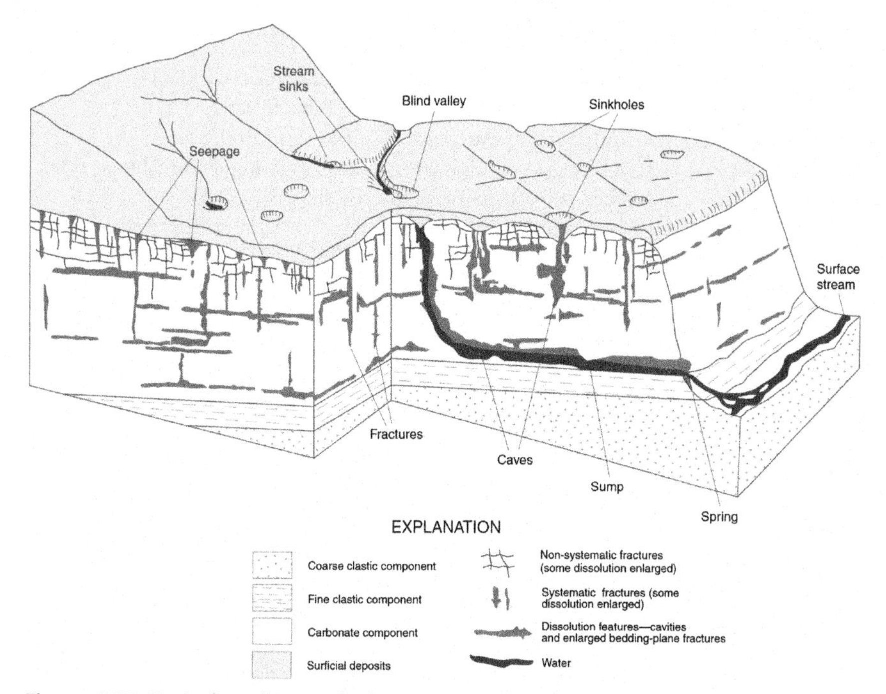

Figure 3.27 Typical attributes of a karst system (Runkel et al., 2003).

investigation is crucial in underground engineering projects as it is essential for a project to be successful.

3.2.1 Scopes of Site Investigation

In the 1968 Rankine Lecture (The Rankine Lecture is one of the most influential lectures in the field of geotechnical engineering. It is named after W. J. M. Rankine, an early contributor to the theory of soil mechanics, and is organized annually by the British Geotechnical Association.) Rudolph Glossop said: "*If you do not know what you should be looking for in a site investigation, you are not likely to find much of value.*" (1968 quoted by Fookes, 1997). As a result, the first step in conducting site investigation is to identify which information is useful and should be acquired.

Site investigation is carried out to collect facts and data that greatly influence planning, design, construction, and operation. From an engineering perspective, the data acquired can be categorized into three broad areas: the geological conditions, hydrogeological conditions, and geotechnical characteristics. The specific contents of each are given in Table 3.7.

Table 3.7 Information to be obtained through site investigation

Geological	Stratigraphy, structure, and identification of principal rock/soil types and their general characteristics
Hydrogeological	Groundwater reservoirs, aquifers, and pressures
Geotechnical	Rock mass/soil characteristics and geomechanical properties
Other	Major water-bearing faults, methane gas, etc.

Source: Data from US Army Corps of Engineers. (1997). *Engineering and design: tunnels and shafts in rock*. EM 1110-2-2901, Department of the US Army, Washington, DC.

However, acquiring only these parameters is far from sufficient for a successful underground engineering project. Other aspects, such as the availability of skilled labor, building materials and construction access, etc., should also be considered during site investigation. The latter are in fact of equal importance to the geological, hydrogeological, and geotechnical data acquired. One must also distinguish between site investigation and ground investigation (Muir Wood, 2000). By definition, the former refers to the total process of determining the nature of the ground relevant to design and construction, and the latter is only part of the former and entails geological and geophysical investigations commissioned solely for the project.

3.2.2 Stages of Site Investigation

When conducting site investigation, it is difficult and inefficient to acquire the complete and comprehensive set of data at once. It is common practice to organize and conduct site investigation in three distinct stages (Kolymbas, 2005):

Stage 1: Preliminary investigation

Stage 2: Main ground investigation

Stage 3: Further investigations undertaken as part of the project

The preliminary investigation is carried out in the preliminary planning stages of a project. In this stage many choices regarding alignments and constructional methods are proposed. In accordance with the project outline, the preliminary investigation serves three major purposes: comparison between alternatives, confirmation of feasibility, and formulation of plans for future investigation. To minimize costs and improve efficiency, preliminary investigation relies on "desk study," which examines existing data from geological maps and evidence from other construction sites. It may also include a walkover survey and selected exploratory borings in critical locations. Through preliminary investigation, general information about topography, geology, and hydrology along the alignment can be obtained or at least estimated.

The main ground investigation is performed during the overall planning, detailed planning, and design stage of a project. At these stages the tunnel alignment has been decided and the main ground investigation is carried out to provide the data needed to facilitate project implementation. It consists of extensive on-site geophysical studies and laboratory tests of core samples extracted along the alignment. The investigation results are interpreted and presented through site investigation reports.

The duration of investigation extends into the tunnel construction stages. Investigations should in fact be carried out continually during the construction in order to update acquired data and attempt to validate predictions. Mappings of the tunnel face and walls are made and numerous measurements of settlements, stresses, and vibrations, just to name a few.

3.2.3 Site Investigation Methods

Site investigation varies from site to site and during different stages. During preliminary investigation, for example, when the project is broadly defined, rapid yet low-cost investigation methods are desired to obtain an overview of the site conditions. But during main ground investigation, the aim is to provide sufficient information for the design and construction stages. The latter therefore requires more rigorous ground exploration. Finally, during the construction stage, relatively simple site investigation is carried out to limit interference with the construction work.

When conducting site investigation, it is advantageous to use existing data. This includes regional geological maps, data from other nearby engineering projects, and even digital information. Since this requires no extra fieldwork, it is particularly suitable for the preliminary investigation phase. One such example is given in Fig. 3.28, which shows a hydrogeological sketch of China.

The site investigation method can be classified as noninvasive and invasive. The noninvasive methods include remote sensing, digital photogrammetry, and geophysical methods, while the invasive methods include

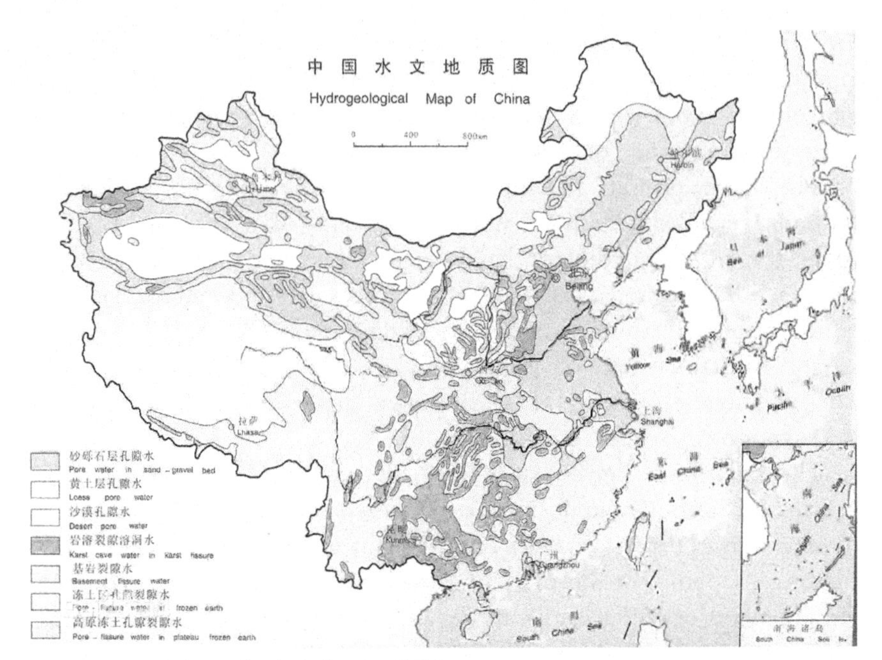

Figure 3.28 Hydrogeological sketch of China.

drilling and sampling methods. Some of the most widely used methods for site investigation will be introduced here, namely remote sensing, field mapping, core-drilling, and several other geophysical methods.

Remote sensing is used to obtain the geological features of the ground surface. Techniques such as air photos can reveal valuable information for field verification such as (US Army Corps of Engineers (US ACE), 1997):

- black-and-white stereo coverage used for topographic mapping, providing information about landform definition, boundaries between rock and soil types, etc.;
- color photos for determining land use; and
- infrared photos showing temperature differences, which are useful for defining drainage paths as well as ground moisture content contrasts.

Having obtained an estimate of the site conditions through desk studies of existing data and aerial photography, the next step is to verify the findings through on-site studies. Common on-site studies begin with field mapping. Engineering geologists visit the project site and observe ground surface geological features, which also provide information about those beneath the surface. For example, the reoccurrence of rock stratigraphy is usually caused by folds. By observing the pattern of rock stratigraphy exposed on the ground surface, this information can be projected to a certain depth.

Core-drilling (Fig. 3.29) is more straightforward than field mapping, but in assessing the geological structures beneath the ground surface. Vertical core drillings are usually placed at either side of the planned tunnel route. The boreholes reach a depth of one tunnel diameter below the tunnel invert. Of course, this is simpler for shallower tunnels than deeper ones. For deep tunnels such as those passing through mountains, the exploration of rock conditions relies more on remote sensing and field mapping. Long horizontal boring from the portals can also be employed. The core-drilling diameter is at least 100 mm (Kolymbas, 2005). Cores with a larger diameter can yield high-quality results, since the small scale in homogeneity is averaged. The cost and labor associated with extracting larger cores are also higher.

It is still not possible to anticipate all unfavorable conditions that might be encountered during a project, such as cavities or water-bearing joints. Geophysical methods should therefore be adopted for more comprehensive exploration of ground conditions.

Geophysical methods (Fig. 3.30) such as electrical resistivity are adopted as supplementary approaches to core drillings in the detection of

Figure 3.29 Drilling of sample.

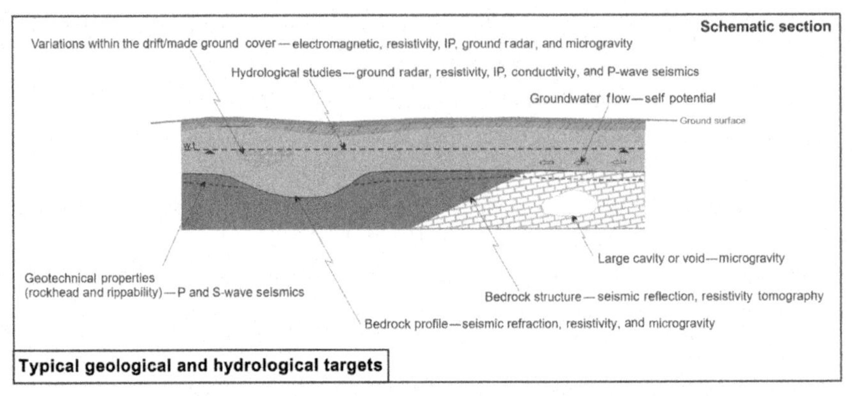

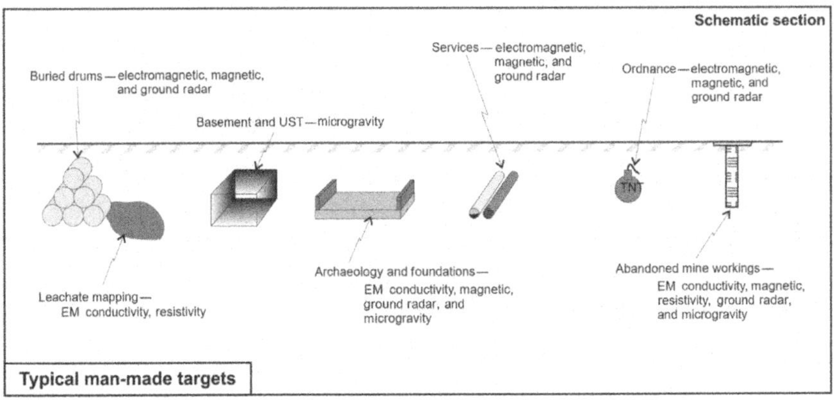

Figure 3.30 Schematic representation of typical natural and man-made targets and the most appropriate geophysical methods for a survey (TerraDat, 2005).

unfavorable conditions. The underlying principle is that ground properties influence the transmission and reflection of seismic, electric, magnetic, and gravitational fields. By measuring the transmission and reflection of certain waves, tunnel engineers can deduce ground properties and characterize different materials. To achieve reliable results, special attention should be paid to the application, surrounding noise, and limitations of the geophysical device employed.

3.2.4 Application of Digital Underground

Underground engineering projects follow a process of continuously acquiring, analyzing, and processing data. This requires permanent access to a range of data throughout all the stages of the project, from initial planning to operation. But today, the challenge is to process and analyze the great quantities of data modern measurement technologies provide. This digitalization must be efficiently managed, as it is a key tool to process, store, and transmit data. It constitutes the basis of exchange between research and applied science.

Underground engineering transitioned into the 21st century through the digitalization of some of the technologies it uses. Some examples include:

- The digital stratum. A technology introduced by Zhu in 1998. It enables to visually show both the primary stratum information (crust movement and surrounding conditions) and the construction disturbance information (human engineering activities) (Zhu, Huang, Li, Zhang, & Liu, 2016).
- The AMADEUS research plan (adaptive and real-time geologic mapping, analysis, and design of underground space). It was an IT-based tunneling system introduced by Virginia Polytechnic university. Its main objective is to integrate IT in improving rock mass characterization, making tunnel design more economical, and improving tunnel construction safety (Gutierrez, Doug, Dove, Mauldon, & Westman, 2006).
- The DUSE (digital underground space and engineering). Proposed in 2007, it is an innovative concept that enables efficient management and monitoring of the investigation process, design, and construction (Fig. 3.31). Inspired by 3D geographic information systems and digitization, the DUSE provides a platform for data sharing and analysis, centralizing engineering construction, maintenance, and disaster protection. Being an efficient way to visualize and manage data, the birth of DUSE has also led to a digital underground engineering museum (Wang, Zhou, & Bai, 2015).

The DUSE system is divided into five levels (data, modeling, representation, analysis, and application) from the perspectives of engineering, information, services, and software systems. The primary function of the data layer is to build a standardized and open database, centralizing all data from underground engineering projects. The modeling layer uses 3D strata and structure modeling methods to construct visualization models of strata, underground structures, and pipelines. Finally, the application layer constitutes a platform where practical problems can be handled,

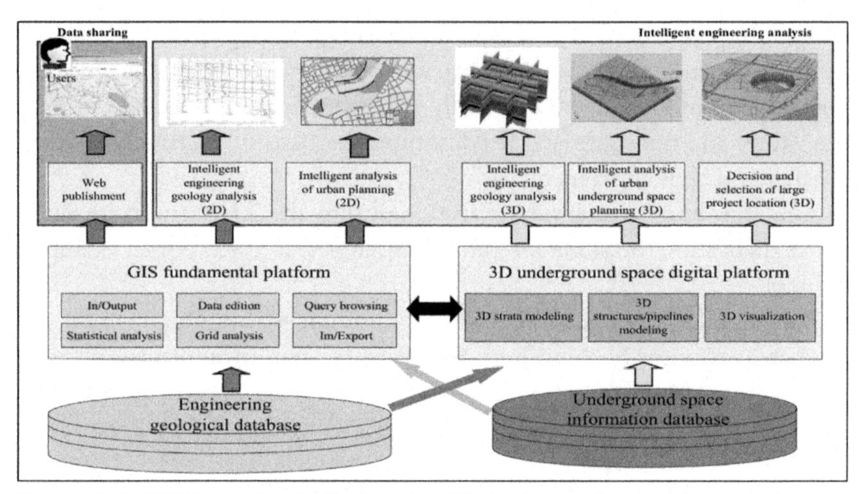

Figure 3.31 DUSE system built for the Changzhou urban underground space resources evaluation project (Zhu et al., 2016).

scientific evaluation carried out, and virtual browsing performed (Zhu et al., 2016).

- The iS3 platform is an integration of software modules and hardware systems. The data acquisition includes a variety of artificial or automated information collection tools, methods, and equipment for engineering objects, and the collected data, graphics, and images and other information are stored in a large-capacity database.

  The original data of a project is generally difficult to be used directly and efficiently. An important concept of iS3 is the standardization of engineering data. Based on data types, corresponding data format standards and data preparation guidelines are formulated and embedded into the iS3 system. It integrates data standards into data acquisition protocols and equipment systems to allow types of data entering the database to have the same format that can be automatically identified and efficiently used. The project involves a huge amount of data, and these data will form an uninterrupted lifecycle of "data flow" in the process of the planning, design, construction, operation, and maintenance of a project. Therefore, big data storage, processing, and analysis are important functions of data representation in iS3.

  iS3 enables a linkage engineering view of geographic information system and 2D/3D. The engineering visualization model is divided into the digital model and the numerical model. The digital model represents the physical objects in the computer with real spatial position and combines the spatial relationship and its attributes in order to realize the one-to-one correspondence between the spatial entity and the attributed information. The numerical model is based on the numerical simulation method and integrates the virtual model, which is the object of the numerical calculation, with corresponding geometric information and mechanical parameters.

Engineering numerical models are generally not directly applicable to numerical analysis. The integration of digitalization and numerical analysis is based on the digital platform and numerical analysis program unifying the engineering digital modeling technology with numerical analysis techniques. These models also standardize and integrate digital and numerical modeling information on a unified 3D engineering model to provide automated and high-precision virtual computing objects. The integration of digital modeling and numerical analysis is harmonized with numerical analysis techniques. The digital and numerical model information is unified for 3D engineering model for standardization and integration. They are also available for engineering design and numerical analysis to provide automated and high-precision virtual computing objects.

iS3 integrates automated 2D and 3D model recognition and transformation algorithms, such as the automatic transformation of geometric point cloud patch model, automatic grid meshing, face recognition and disk-based, space cutting algorithm, etc. It can automatically transfer a digital model into a numerical model, or integrate digital model information with numerical model information into the same model. In this way, the visualization models and data can be used to support digital project management similar to BIM and be directly used for numerical analysis.

3.2.5 Site Investigation Reports

Site investigation reports are documentation and representation of the site investigation results. They are used to transfer information from engineering geologists to design and construction teams. These reports are made regularly during site investigation. A summary is usually required at the

Table 3.8 General layout of a GBR

Items and subitems
1. *Introduction* Project name, project client, purpose of the report, etc.
2. *Project description* Project location, project type and purpose, summary of key project features, etc.
3. *Sources of geologic information* Reference to the geotechnical data report (GDR), etc.
4. *Project geologic setting* Brief overview of geologic and groundwater setting, etc.
5. *Project construction experience* Nearby relevant projects, summary of construction problems and how they were overcome, etc.
6. *Ground characterization* Physical characteristics, ranges and values for baseline purposes, laboratory and field test results, etc.
7. *Design considerations* Criteria and methodologies used for design, environmental performance considerations, etc.
8. *Construction considerations—tunnels and shafts* Anticipated ground behavior, specific anticipated construction difficulties, etc.

Source: Adapted from Essex, J. R. (1997). *Geotechnical baseline reports for underground construction: guidelines and practices*. Reston, VA: American Society of Civil Engineers.

end of preliminary investigation and reporting necessary after the completion of each phase of the main ground investigation.

The site investigation results also have contractual use, meaning the client of an underground engineering project must provide sufficient investigation data for bidding purposes. This is referred as a Geotechnical Baseline Report (GBR) and serves as part of the contract documents and so is more formal in structure and content than regular investigation reports. Table 3.8 shows the typical layout of a GBR.

In the GBR, the geological, hydrological, and geotechnical conditions are shown to bidders. This includes undesired effects that may occur, such as the presence of harmful gases, active joints, dangerous substances, etc. Such information must be provided to bidders, though it cannot all be provided due to the volume and complexity of all preoperation investigations. The minimum to be included must be the boring logs, test trenches, and adit data (US ACE, 1997).

When formulating a report, one must also distinguish between raw and interpreted data. In design-build contracts (where the designer and builder are the same), the client of the project should not present interpretational conclusions that might influence construction methods. In other cases, where the design and construction processes are separated, the designer can include a geotechnical interpretative report (GIR) in contract documents. It is still a geotechnical data report (not a design scheme) but contains predictions and estimations of ground conditions. The GIR needs to be carefully written and reviewed, because it is the baseline against which the construction team can make claims for different site conditions.

3.3 INSTRUMENTATION AND MONITORING

Instrumentation and monitoring techniques are adopted in almost all fields of civil engineering. Accelerometers, for instance, are used on buildings to record their seismic performance, and infrared imaging is used to verify the integrity of dams and pipelines. Even strain gauges used in laboratory experimentation are a form of instrumentation.

Instrumentation technologies are sophisticated today. Prefabricated into delicate and highly integrated modules, many are palm-sized. There exist many different devices for many different types of instrumentation (Fig. 3.32).

Instrumentation and monitoring in underground engineering are of paramount importance and thus should be controlled accordingly. Engineers are responsible for the choice of parameters to be monitored, the instrumentation techniques to be used, and how the results they provide are to be interpreted. For example, they should decide whether parameters should be obtained through continuous real-time monitoring or not.

3.3.1 Purposes of Instrumentation and Monitoring

Instrumentation and monitoring in underground engineering, in particular for tunneling projects, serves two purposes:

1. to provide information pertaining to imminent collapse, thus making necessary countermeasures possible and
2. to act as a feedback mechanism, enabling continuous adaptation of design when predictions deviate from measured values

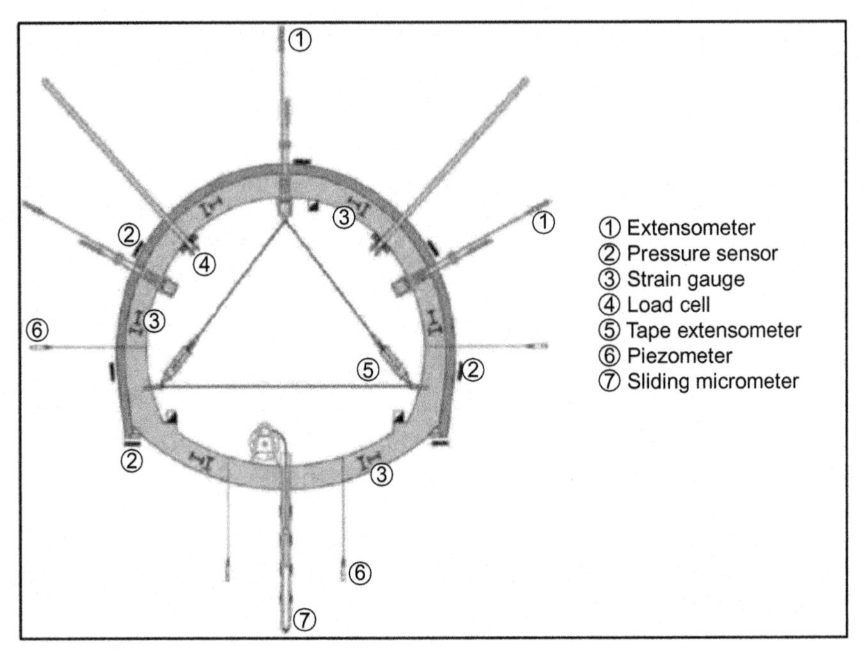

Figure 3.32 Typical instrumentation layout. ITA-CET (2009). Tunneling Course Material-Tunneling in Hot Climate Country, Monitoring of Tunnels, Riyadh, *reproduced by ITA Working Group n°2. (2011).* Monitoring and control in tunnel construction. *ITA Report 9. France, Avignon.*

The second purpose is often neglected. However, the role of instrumentation and monitoring in a feedback design scheme is crucial. Constructing a tunnel that will not collapse is indeed insufficient for underground engineers. The project also needs to be economical and considerate of the local environment. This is best realized through feedback analysis.

More precisely, the main objectives of instrumentation and monitoring of underground projects are to (ITA WG2, 2011):

- obtain information regarding ground response to tunneling;
- provide construction control;
- verify design parameters and models;
- measure lining performances during and after construction;
- monitor impacts on surrounding environments such as ground settlement and groundwater regime;
- warn of any safety-critical trends;

- predict future trends in monitored and not yet monitored parameters; and
- make predictions on the performance and management of completed tunnels.

3.3.2 Parameters to Be Monitored

Choosing the appropriate parameters to be monitored is necessary for a successful monitoring scheme. The appropriate parameters can provide direct answers to problems engineers face. For example, if engineers want to know whether the number of rock bolts is adequate, the parameter of interest is the load each is subjected to. Monitoring tunnel wall displacements can also provide useful information for this problem. Table 3.9 gives a list of parameters that often need to be monitored.

As Table 3.10 shows, leading parameters to be monitored can differ depending on the construction method employed.

3.3.3 Types of Instrumentation

Having identified the leading parameters to be monitored, it is possible to find suitable instrumentation for monitoring. The equipment can be categorized into groups based on the parameters it can measure. Table 3.11 shows some commonly used devices.

When selecting an instrumentation method, one must consider range, accuracy, and precision requirements. The instruments should also be

Table 3.9 Typical monitored parameters

Location	Parameter
Tunnels, underground chambers, shafts, and portals	Convergence, crown settlement, floor heave, distribution of deformation behind the rock wall, load in dowels and anchors, stress in the lining, groundwater pressure within the rock mass, water pressure acting on the lining
Urban environments	Surface settlement, vertical and horizontal deformation of buildings and other structures, vertical and horizontal deformation of the ground at depth groundwater pressure

Source: Adapted from US Army Corps of Engineers. (1997). *Engineering and design: tunnels and shafts in rock*. EM 1110-2-2901, Department of the US Army, Washington, DC.

Table 3.10 Leading parameters to be monitored

	Conventional method (low overburden in urban areas)	**Closed-face TBM (urban areas)**	**Operational tunnel in creeping ground**
Visual inspection	●	○	●
Geometrical parameters			
Face extrusion	●		
Surface settlement	●	●	
Surface rotation		○	
Extrusion of the ground ahead of tunnel face	■		
Displacement in borehole	○	○	×
Convergence at sidewall	●		●
Crack monitoring	○	○	●
Deformation of permanent lining	×	×	●
Mechanical parameters			
Force (arch base, anchoring rod, rock bolt, etc.)	●		
Stress in ground			○
Stress in support/lining	○	×	○
Hydraulic parameters			
Pumped out water rate	○		○
Surface rainfall	×		×
Piezometric levels in ground	●	●	●
Temperature of leakage			×
Other parameters			
Tunnel air temperature		○	×
Tunnel air pressure		×	
Tunnel hygrometry	×	×	○
Date and time	●	●	●
Vibration from blasting	●		

(×) Usually secondary parameter; (○) frequently important parameter; (●) essential parameter, always monitored; (■) essential parameter, always monitored with advance full-face preconfinement.
Source: Adapted and modified from AFTES Groupe de Travail n°19. (2005). *Guidelines on monitoring methods for underground engineering*. Tunnels et ouvrages souterrains, 187, 11–21 (in French); translated by International Tunnelling Association (ITA) Working Group n°2. (2011). *Monitoring and control in tunnel construction*. ITA Report 9. France, Avignon.

Table 3.11 Typical instrumentation equipment

Parameter	Equipment
Settlement at the surface	Level
Displacement and convergences in the opening skin	Tachymeter, tunnel scanner, extensive tape, etc.
Displacements within the rock mass	Extensometers, sliding micrometers, inclinometers, deflectometers
Stresses on support	Pressure cells, strain gauges
Strains on support	Strain gauges
Primary or natural stresses in the ground	Unloading drills

Source: Adapted from ITA. (n.d.). *How to go underground—Design*. Retrieved from <http://tunnel.ita-aites.org/en/how-to-go-undergound>.

Table 3.12 Comparison between two strain-measure instruments

Instrumentation	Range	Resolution	Accuracy
Vibrating wire strain gauges	Up to 3000 με	0.5–1.0 με	± 1–4 με
Fiber optics	To 10,000 με (1% strain)	5 με	20 με

Source: Adapted from The British Tunnelling Society, & The Institution of Civil Engineers. (2004). *Tunnel lining design guide*. London, England: Thomas Telford.

calibrated before use. As an example, the range, resolution, and accuracy of strain gauges and fiber optics for use in a strain measurement are shown in Table 3.12. Though strain gauges seem to be a better choice than fiber optics in this case, the latter also has other advantages. Depending on the application, different instruments are suitable.

3.3.4 Interpreting Monitoring Results

After obtaining monitoring results, the next step is to interpret and use them for tunneling control. Instrumentation and monitoring of tunneling projects yields vast amounts of useful information, such as convergence, absolute displacement, extrusion and stress levels. In this section though, only convergence measurements are taken as an example to illustrate the concept of results interpretation.

Tunnel convergence refers to the change of distance between two points on the same tunneling cross section over time. It may be attributed to face advance and time-dependent behaviors of the rock mass. It is an ideal parameter to evaluate tunneling activities since it is an integrated quantity that reflects major local effects. Other parameters such as stress, strains, and curvatures are significantly influenced by local effects. As a result, convergence measurements provide a good overall assessment of tunneling activities.

Convergence at time t between any two opposite points in a tunnel cross section $c(t)$ is defined as

$$c(t) = D_0 - D(t) \tag{3.20}$$

where D_0 is the initial measurement of the distance between these two points, and $D(t)$ is the measurement at time t.

A typical convergence instrumentation scheme is shown in Fig. 3.33 where points 1–4 are pins or target plates mounted onto the tunnel wall immediately after excavation. The typical method is to measure the distance between the pins using optical methods.

For convergence analysis, the measured data is plotted on three graphs (Fig. 3.34):

1. Distance x (between convergence station and face excavation) versus time t. This reflects the progress of face advancement as well as the advancing speed.
2. Convergence c versus time t. This is used to evaluate the face influence.
3. Convergence c versus distance x. This is also used to evaluate the face influence.

Of the three graphs shown in Fig. 3.34, graph (A) shows that tunneling activity was interrupted for a first time during the time range t_2–t_3

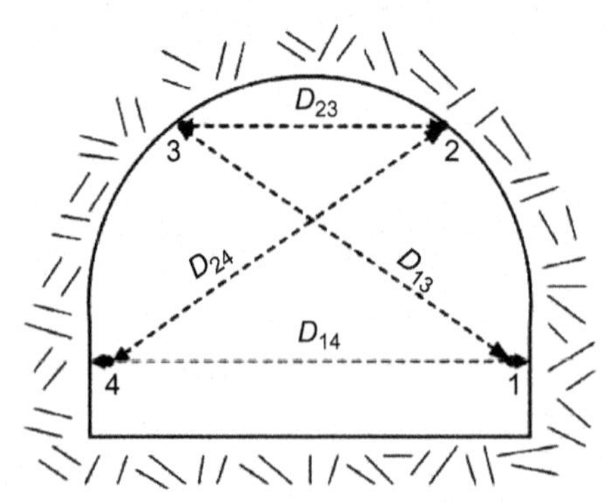

Figure 3.33 The convergence measurement scheme. *Hudson (1995), reproduced by ITA Working Group n°2. (2011).* Monitoring and control in tunnel construction. *ITA Report 9. France, Avignon.*

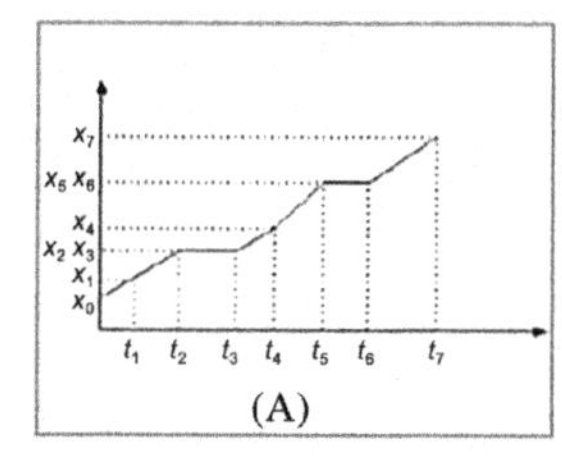

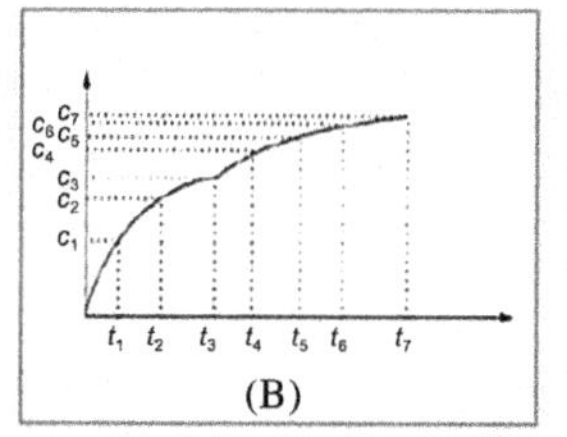

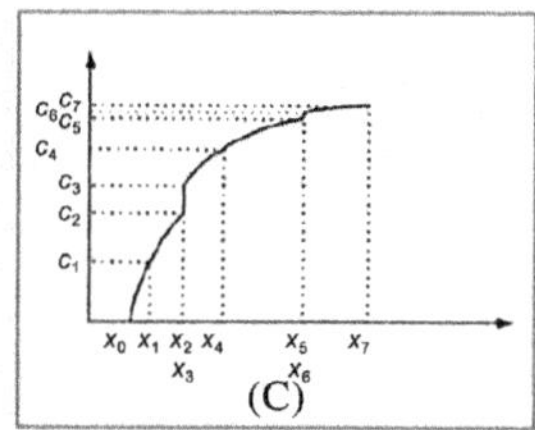

Figure 3.34 Graphs for convergence analysis. (A) interruption of tunnel advance for t_2-t_3 and t_6-t_5, (B) discontinuity due to resumed face advance and (c) convergence versus face distance. *Hudson (1995), reproduced by ITA Working Group n°2. (2011).* Monitoring and control in tunnel construction. *ITA Report 9. France, Avignon.*

(distance x_2, x_3) and a second time during t_5-t_6 (distance x_5, x_6). During these interruption periods, if the convergence value c increases, the time-dependent ground mass properties (e.g., creep) are involved as graph (B) shows. Moreover, as the tunneling face advances and the distance x increases, the influence caused by face excavation tends to decrease. Graph (B) reflects this, showing that the convergence value increase is less significant during the second interruption compared to the first one. The slopes of graphs (B) and (C) are therefore important indicators of the tunnel stability. In the example of Fig. 3.34, the tunnel is stable as the slopes approach zero with an increase of time t and distance x, respectively.

3.4 BACK ANALYSIS

3.4.1 Definition

Underground engineering deals with complicated and unforeseeable construction environments. Site investigation and laboratory tests alone cannot eliminate all the uncertainties. To control and overcome these ground-related uncertainties, tunnel engineers (in particular Terzahi and Peck) have put forward the observational method. It involves closely monitoring of the construction process to estimate the stability of the underground structures and verify and/or update the initial design and construction method.

The first role was discussed in Section 3.4.4 with the example of convergence monitoring. Back analysis corresponds to the second role—using field measurements to obtain input parameters. Back analysis has become an effective tool in tunneling. Its advantage is that it can reduce design parameter deviations from the actual conditions to a minimum (Gioda & Sakkurai, 1987). But back analysis also has drawbacks, such as the nonuniqueness of the solution. For instance, the results it provides that depend on the selection of measured quantities can lead to different back analysis solutions (Sakurai, Akutagawa, Takeuchi, Shinji, & Shimizu, 2003).

3.4.2 Process

Fig. 3.35 depicts the scheme of the overall design and construction process incorporated in back analysis. To clearly appreciate this flow chart, the steps are divided into two parts. Part I, from preliminary research to construction, is a standard forward analysis scheme. Material properties and other parameters are estimated through site investigations or laboratory tests. Design and construction is then are carried out based on these parameters (which may not reflect the real situation due to the accuracy of the tests, etc.).

This forward analysis scheme works well for aboveground structures. However, for underground projects, due to complex and unforeseeable field conditions, actual performance may deviate greatly from the desired outcomes. Part II therefore follows from the end of Part I. During the construction stage physical quantities are closely monitored such as displacements as they are less influenced by local effects and easy to measure. This monitored data is processed and used in a back analysis procedure to determine the unknown parameters or to update the original model in the initial design.

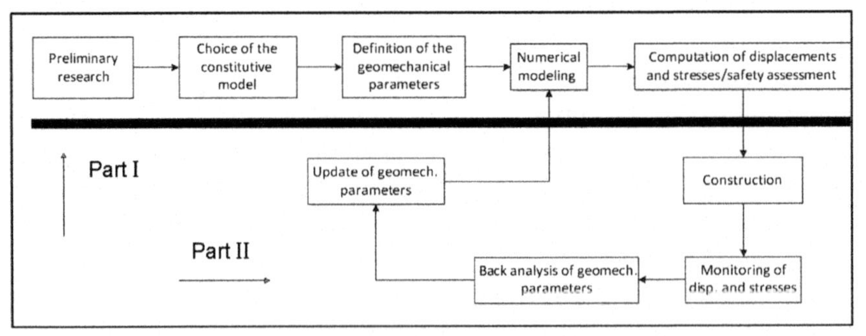

Figure 3.35 Scheme of the observational method using back analysis in the process. *Modified from Miranda, T., Dias, D., Eclaircy-Caudron, S., Gomes Correia, A., & Costa, L. (2011). Back analysis of geomechanical parameters by optimisation of a 3D model of an underground structure.* Tunnelling and Underground Space Technology, *26(6), 659–673.*

There are two distinct methods to conduct back analysis: the inverse and direct approaches. The former is to inversely write governing equations so that material parameters appear as the outputs and the measured quantities as inputs. This is an efficient approach, though modifying software to rewrite the equations is impossible.

Contrary to the inverse approach, the direct approach adopts a trial-and-error process. An error function is first established to calculate the differences between the observed and computed quantities. The design parameters are then modified iteratively until convergence to the global minimum error is achieved. This approach requires much iteration and is therefore more time-consuming than the inverse method. But the direct approach is still widely employed, as it offers the advantage of greater flexibility as its error function can be selected independently from the numerical model and coupled to the design routine through simple programming (Miranda et al., 2011).

3.5 CASE STUDY: PLANNING AND DESIGN OF THE SHANGHAI YANGTZE RIVER TUNNEL

The Shanghai Yangtze River Tunnel and Bridge project is located at the South Channel waterway and North Channel waterway of the Yangtze River mouth in the northeast of Shanghai, which is a significant part of the national expressway, as shown in Fig. 3.36. It is an extremely major transport infrastructure project at seashore in China at Yangtze River mouth and also the largest tunnel and bridge combination project worldwide. The completion of the project will further promote the development space for Shanghai, improve the structure and layout of the Shanghai traffic system, develop resources on Chongming Island, accelerate economic development in the north of Jiangsu Province, increase the economy capacity of Pudong, accelerate the economy integrity of Yangtze River Delta, and create a boom to the economic development of the Yangtze River area and even the whole country and upgrade the comprehensive competence of Shanghai in China and beyond.

The Shanghai Yangtze River Tunnel and Bridge project alignment solution was first implemented based on Shanghai urban planning. The project consist on a planned western solution, which was decided as an optimal solution after prior studies involving various aspects and compared with the east alignment. The western alignment starts from Wuhaogou in Pudong, crossing Yangtze River South Channel waterway

Figure 3.36 Location of Chongming Crossing.

to Changxing Island and spanning the Yangtze River North Channel waterway to the east of Chongming Island.

The Yangtze River begins is divided into three levels of branches and with four mouths flowing into the sea: The South Channel waterway is a mixed river trench. The intermediate slow flow area forms Ruifeng shoal, which is relatively stable for a long time. The natural water depth makes it the main navigation channel. However, the North Channel waterway is located in the middle part of the river, which is influenced by the south part and branch transition into the North Channel waterway. So the trench varies alternatively and the river map is not as stable as the South Channel waterway. Therefore, after iterative discussion by several parties, the Southern Tunnel & Northern Bridge 5 solution is selected. The total project is 25.5 km long, among which 8.95 km is tunnel with a design speed of 80 km/h; 9.97 km is bridge; and 6.58 km is land connection, with a design speed of 100 km/h, as shown in Fig. 3.36. The total roadway is planned as dual six lanes.

3.5.1 Three Methods of Tunnel Lining Design Given by the ITA

In this section, three tunnel lining design methods are discussed. Subsequent to the calculation of the specific conditions, the internal forces of the sections are compared to provide reference for the selection of engineering design methods.

3.5.1.1 Design Methods for Tunnel Lining

In this section, three commonly used lining design methods are reviewed as follows:

1. *Bedded frame model method*

   The bedded frame model method is a method used to compute member forces of tunnel lining with a matrix method. This method can evaluate the following conditions (ITA, 2000):
   - nonuniformly varying load due to change of soil condition;
   - eccentric loads;
   - hydrostatic pressure;
   - spring force to simulate subgrade reaction; and
   - effect of joint by simulating joints as hinges or rotation springs.

   A typical adaptable loading model for the bedded frame method is shown in Fig. 3.37.

2. *Elastic equation method*

   The elastic equation method, which is commonly used in China and Japan, is a simple method for calculating member forces without a computer. However, it cannot evaluate the conditions mentioned earlier. In this method, water pressure is evaluated as the combination of vertical uniform load and horizontally uniformly varying load. The horizontal subgrade reaction is simplified as a triangularly varying load.

3. *Muir wood model*

   Muir Wood (2000) gave a corrected version of Morgan's more intuitive approach, again assuming an elliptical deformation mode (Fig. 3.38). The tangential ground stresses are included, but radial deformation, which is due to the tangential stresses, is omitted. Making allowances for some predecompression of the ground around the opening before placing the lining, Muir Wood considers only 50% of the initial ground stresses. The moments may be reduced even more by reducing the lining stiffness by an amount equivalent to the effect of less rigid hinges.

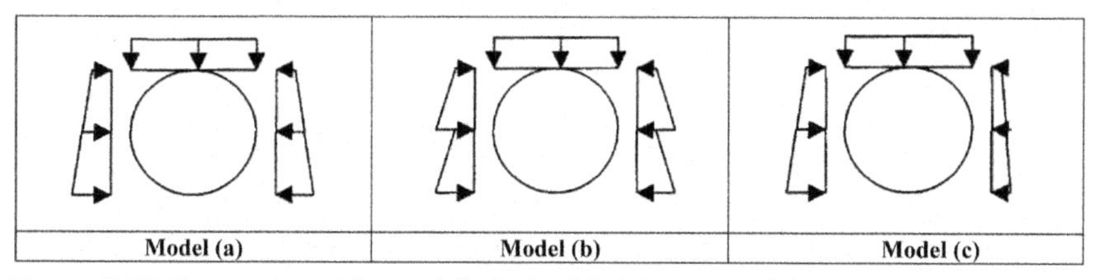

Figure 3.37 Three adaptable models for bedded frame model.

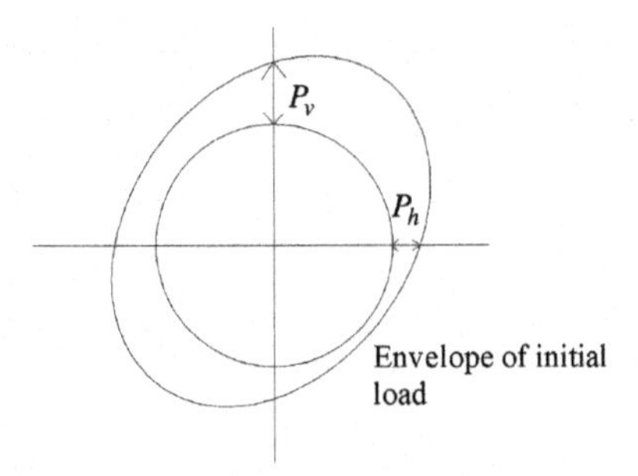

Figure 3.38 Initial loading on tunnel prior to deformation.

Based on the Muir Wood model, the internal forces follow these equations:

$$N_{Hoop} = ((P_{V\max} + P_{H\max})/(2(1+RC)))R_0$$
$$RC = R_0 E_{Soil}(1-\nu_1^2)/\eta t E_{Conc}(1+\nu)$$
$$I_e = I_j + I_s(4/n_s)$$
$$N_{\max} = N_{Hoop} + M_{\max}/R_0$$
$$N_{\max} = N_{Hoop} - M_{\max}/R_0$$
$$M_{\max} = (P_{V\max} - P_{H\max})(\eta R_0)2/\left[6+2\left\{(\eta R_0)3E_{Soil}/[E_{Conc}I_e(1-\nu)(5-6\nu)]\right\}\right]$$

where

N_{Hoop}, circumferential hoop thrust per unit length of tunnel; N_{Max}, maximum axial force in lining per unit length of tunnel; N_{Min}, minimum axial force in lining per unit length of tunnel; M_{Max}, maximum bending moment in lining per unit length of tunnel; P_{VMax}, maximum vertical loading acting on tunnel crown P_{HMax}, maximum horizontal loading acting on horizontal axis of tunnel; R_0, radius to extrados of inner lining; ν_1, Poisson's ratio for lining; ν = Poisson's ratio for ground; E_{Soil}, Young's modulus for ground; E_{Conc}, Young's modulus for lining; T, segment thickness; η, ratio of radius of lining centroid to that of extrados; n_S, number of segments per ring; I_S, second moment of area of lining per unit length of tunnel; I_j, second moment of area of joint; and I_e, effective value of second moment of area for a jointed lining.

3.5.1.2 Comparison Between Tunnel Lining Design Methods

The abovementioned tunnel lining design methods are used to compute bending moment and axial force. However, member forces calculated by different methods are not consistent, varying results along design assumptions. For the Chongming Tunnel Project a number of loading cases was used to calculate member forces. In accordance with the analysis results, the limitations and suitability of such design methods were evaluated.

1. *Bending moment*

   The maximum bending moment computed by different design methods at each loading case show that in all loading cases, the maximum and minimum bending moments were obtained by the elastic equation method and the Muir Wood model, respectively. The bending moment computed by the elastic equation is comparatively conservative, whereas Muir Wood is immoderate. It is important to note that such overestimation of bending moment will induce high costs, whereas high risk of tunnel instability will be induced by underestimation of bending.

2. *Axial force*

   The maximum and minimum axial forces computed by different design methods for each loading case show that the difference of axial force between each method is not significant. From the bending moment values results, it is shown that the axial force will be direct dependent on the soil cover and water depth above the tunnel crown.

The reader can refer to the reference Frew (2008) for further information and specific calculation data.

3.6 QUESTIONS

3.1. What are the names of these two platform types? What are their advantages and disadvantages?

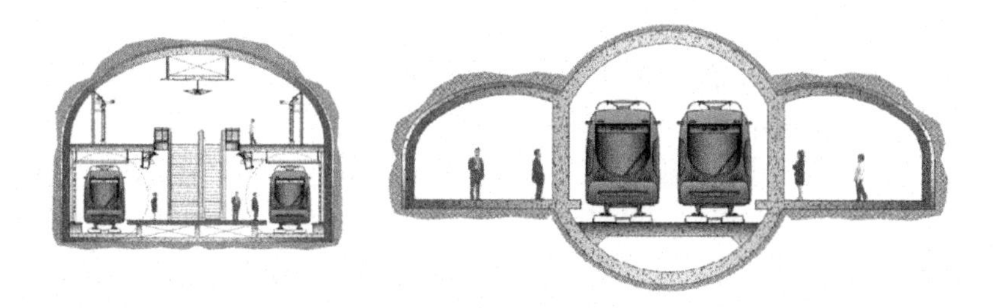

3.2. Give the full names of the following acronyms and indicate which methods are generally used for soil and rock.

1. FEM

2. DEM
3. BEM
4. DFN

3.3. Conventional tunneling allows either the full-face or partial excavation of a tunnel cross section according to the rock class. Choose the suitable excavation method for the following rock classes. ((1) central diaphragm excavation; (2) full-face excavation; (3) short stage method; (4) sidewall adit heading excavation; (5) crown heading excavation.)

- Classes I–III in terms of small-sized tunnels ( )
- Class IV and V, while the bench length is 10–15 m ( )
- Class IV or V, typically useful in fragmentary strata ( )
- Class IV and V, with a temporary support in the middle ( )
- Class V, with shallow depth in clay and silt layers ( )

3.4. It is known that lining segments are assembled in the process of shield tunneling (Fig. 1). Is it possible to not use assembly (i.e., employing an integral tube like pipe jacking) in the shield tunnel to complete the work? How could be it done? (Fig. 2)

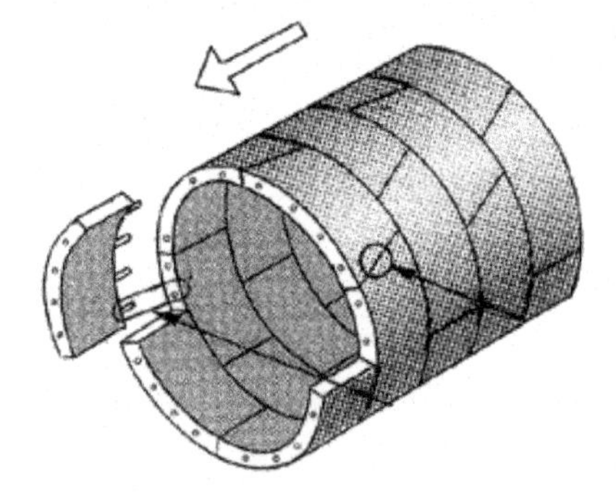

Figure 1 Lining segments.

Figure 2 Pipe.

3.5. Why is geotechnical investigation so important in underground engineering?

3.6. Look at the diagram below and explain why the support in the NATM tunnel is flexible and was not constructed immediately after the excavation. (Note that *Pi* is the required support resistance and ΔR is the radial displacement.)

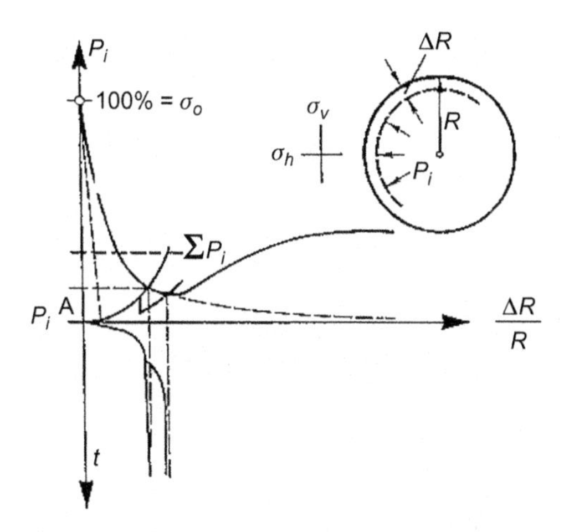

3.7. In an article on geological and geotechnical investigations for tunneling is estimated that "even comprehensive exploration programs recover a relatively mini-scale drill core volume, less than 0.0005% of the excavated volume of the tunnel." Do you think that sampling only this proportion of the rock mass is enough?

3.8. What are the factors that affect the selection criteria for tunneling method machine?

3.9. It is said that the design of underground structures is different from aboveground structures, because of the complicated geological conditions and the structure–ground interactions. The site investigation is crucial in underground engineering, thus reports are generated on a regular basis during site investigation. Explain the meaning and the differences between the geotechnical baseline report and the geotechnical interpretive report.

REFERENCES

Abbas, S. M., & Konietzky, H. (2015). Rock mass classification systems. In H. Konietzky (Ed.), *Introduction to geomechanics*. Freiberg, Germany: TU Bergakademie Freiberg.

AFTES Groupe de Travail n°19. (2005). Guidelines on monitoring methods for underground engineering. *Tunnels et ouvrages souterrains*, *187*, 11–21.

Barton, N., Lien, R., & Lunde, J. (1974). Engineering classification of rock masses for the design of tunnel support. *Rock Mechanics*, *6*(4), 189–236.

Bienawski, Z. T. (1989). *Engineering rock mass classifications: A complete manual for engineers and geologists in mining, civil, and petroleum engineering*. New York: Wiley.

BuildingSmart: Industry Foundation Classes (IFC) – The building SMART data model. www.buildingsmart.org/standards/ifc. Accessed 10.15.

Deere, D. U., & Deere, D. W. (1988). The Rock Quality Designation (RQD) index in practice. In L. Kirkaldie (Ed.), *Rock classification system for engineering purposes* (pp. 91–101). Philadelphia, PA: American Society for Testing and Materials, ASTM STP 984.

Essex, J. R. (1997). *Geotechnical baseline reports for underground construction: Guidelines and practices*. Reston, VA: American Society of Civil Engineers.

Fookes, P. G. (1997). Geology for engineers: The geological model, prediction and performance. *Quarterly Journal of Engineering Geology, 30*, 293–424.

Frew, B. (2008). A review of shield tunnel lining design. *The Shanghai Yangtze River tunnel theory, design and construction*. Taylor & Francis, London, New York. p. 5.

Gioda, G., & Sakkurai, S. (1987). Back analysis procedures for the interpretation of field measurements in geomechanics. *International Journal for Numerical and Analytical Methods in Geomechanics, 11*, 555–583.

Gutierrez, M., Doug, B., Dove, J., Mauldon, M., & Westman, E. (2006). An IT-based system for planning, designing and constructing tunnels in rocks. *Tunnelling and Underground Space Technology, 21*(3–4), 221.

Hudson, J. A. (1995). *Comprehensive rock engineering, vol. 4: Excavation, support and monitoring* (p. 849) Pergamon.

Hudson, J. A., & Harrison, J. P. (1997). *Engineering rock mechanics: An introduction to the principles* (1st ed). Oxford, UK: Pergamon.

ITA. (n.d.). *How to go underground—design*. Retrieved from <http://tunnel.ita-aites.org/en/how-to-go-undergound>.

ITA Working Group n°2. (2000). Guidelines for the design of shield tunnel lining. *Tunnelling and Underground Space Technology, 15*(3), 303–331.

ITA-CET (2009). *Tunneling Course Material-Tunneling in Hot Climate Country, Monitoring of Tunnels*. Riyadh.

ITA Working Group n°2. (2011). *Monitoring and control in tunnel construction*. ITA Report 9. France, Avignon.

Kolymbas, D. (2005). *Tunnelling and tunnel mechanics: A rational approach to tunnelling* (1st ed.). Germany: Springer.

Mana, A. I., & Clough, G. W. (1981). Prediction of movements for braced cuts in clay. *Journal of Geotechnical Engineering, 107*, 759–777, American Society of Civil Engineers.

Miranda, T., Dias, D., Eclaircy-Caudron, S., Gomes Correia, A., & Costa, L. (2011). Back analysis of geomechanical parameters by optimisation of a 3D model of an underground structure. *Tunnelling and Underground Space Technology, 26*(6), 659–673.

Muir Wood, A. (2000). *Tunnelling: Management by design*. London, England: E & FN SPON.

NGI. (2015). *Using the Q-system: Rock mass classification and support design*. Oslo, Norway: NGI.

Palmstrom, A., & Broch, E. (2006). Use and misuse of rock mass classification systems with particular reference to the Q-system. *Tunnels and Underground Space Technology, 21*, 575–593.

Panet, M., Bouvard, A., Dardard, B., Dubois, P., Givet, O., Guilloux, A., & Wong, H. (2001). *The convergence-confinement method*. AFTES.

Runkel, A. C., Tipping, R. G., Alexander, E. C., Jr., Green, J. A., Mossler, J. H., & Alexander, S. C. (2003). Hydrogeology of the paleozoic bedrock in Southeastern Minnesota. *Report of Investigations 61*. Minnesota Geological Survey.

Sakurai, S., Akutagawa, S., Takeuchi, K., Shinji, M., & Shimizu, N. (2003). Back analysis for tunnel engineering as a modern observational method. *Tunnelling and Underground Space, 18*(2–3), 185–196.

Sharma, K. G. (2009). Numerical analysis of underground structures. *Indian Geotechnical Journal, 39*(1), 1–63.

TerraDat. (2005). *A background to shallow geophysical methods with applied engineering & environmental case studies*. Cardiff, UK: TerraDat.

Terzaghi, K., & Peck, R. B. (1967). *Soil mechanics in engineering practice* (2nd ed., p. 729) New York: John Wiley & Sons.

The British Tunnelling Society., & The Institution of Civil Engineers. (2004). *Tunnel lining design guide*. London, England: Thomas Telford.

US Army Corps of Engineers. (1997). *Engineering and design: Tunnels and shafts in rock*. Washington, DC: EM 1110-2-2901, Department of the US Army.

Wang, G., Zhou, Z. Y., & Bai, Y. F. (2015). Digitization and application research of the shield tunnel lifecycle information. *China Civil Engineering Journal, 48*(5), 107–114. (in Chinese).

Zhang, C. P., Han, K. H., Fang, Q., & Zhang, D. L. (2014). Functional catastrophe analysis of collapse mechanisms for deep tunnels based on the Hoek–Brown failure criterion. *Journal of Zhejiang University-Science A (Applied Physics & Engineering), 15*(9), 723–731.

Zhu, H. H., Huang, X. B., Li, X. J., Zhang, L. Y., & Liu, X. Z. (2016). Evaluation of urban underground space resources usingdigitalization technologies. *Underground Space, 1*(2), 124–136.

FURTHER READING

Ding, W. Q., Yue, Z. Q., Tham, L. G., et al. (2004). Analysis of shield tunnel. *International Journal for Numerical and Analytical Methods in Geomechanics, 28*, 57–91.

Frew, B., Wong, K. F., Du, F., & Mok, C. K. (2008). A review of shield tunnel lining design. *The Shanghai Yangtze River Tunnel Theory, Design and Construction*. Taylor & Francis, London, New York, p. 5.

G.C.O. Publication No. 1/90, "Review of Design Methods for Excavations".

Geological Team. (No. 327), 2007. *Bureau of geology and mineral resources of anhui province*. Retrieved from <http://d.ahwmw.cn/szjggw/ahdk327/?a=show&id=291591&m=article>.

McQuail, D. (2010). *McQuail's mass communication theory* (6th ed). London, UK: Sage Publications.

Peck, R. B. (1969). Deep excavations and tunnelling in soft ground. In *Seventh international conference on soil mechanics and foundation engineering—state of the art volume*, Mexico.

CHAPTER 4

Underground Construction

Contents

Underground Engineering
DOI: https://doi.org/10.1016/B978-0-12-812702-5.00004-9

For underground structures the construction method is selected according to geology, geotechnical characteristics, topography, environmental conditions, and underground structure size and depth. Until recently there were few construction methods available due to geological, environmental, economical, and anthropogenic constraints. For shallow structures, a foundation pit or trench is dug from the surface before the lining is built. This is known as the cut-and-cover method. For deeper structures under land, subsurface excavation methods are adopted rather than digging from the surface. This can either be conventional or mechanized tunneling. For underground water structures, an immersed tube method can also be used.

4.1 CUT-AND-COVER CONSTRUCTION

Cut-and-cover tunneling (also known as surface tunneling or open excavation) involves excavating a trench, building an underground structure, and then backfilling and restoring the original ground. This method is used to build shallow tunnels (less than 30 m in depth) and is especially suitable for building metro stations. Examples include the Chengshousi metro station in Beijing, the Siping Road metro station in Shanghai, and the Qianhai metro station in Shenzhen. The technique is widely used, as it is relatively cheap, simple, and easy to implement.

There are three variations of the cut-and-cover method: the bottom-up method, top-down method, and semi-top-down method. Other methods that are adopted in soft and hard ground will be developed too.

4.1.1 Bottom-Up Method

With the bottom-up method, a trench is excavated from the surface within which the structure is built. Once the structure is finished, the trench is filled to restore the surface. The construction method can be broken down into several steps (Fig. 4.1):

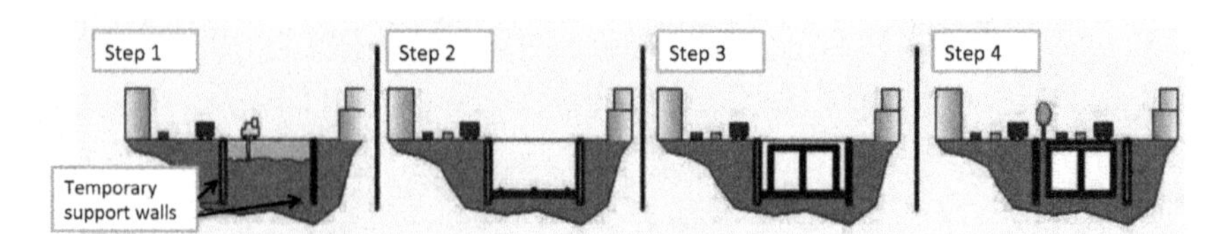

Figure 4.1 Procedure for bottom-up construction.

1. Installation of temporary support walls, excavation and dewatering if required
2. Construction of tunnel by constructing the floor
3. Completion of construction of the walls and roof and waterproofing if required
4. Backfilling to final grade and restoring the ground surface

4.1.2 Top-Down Method

To relieve heavy traffic in urban areas during construction, the top-down method is often employed. After trench excavation, this method involves building a tunnel slab so that aboveground traffic can resume. The method is applied to soft ground and the initial support walls are usually the final structural walls. This form of construction is more complex and time consuming than the bottom-up method, as the working space is more confined.

The top-down construction steps (Fig. 4.2) are as follows:

1. Installation of excavation support/tunnel structural walls and dewatering if required
2. Excavation to bottom level of top structural slab, construction and waterproofing of top tunnel slab
3. Backfilling of roof, ground surface restoring, excavation of tunnel interior, installation of bracing of support of walls, construction of tunnel floor slab
4. Completion of interior finishes including secondary walls

4.1.3 Semi-Top-Down Method

Since the bottom-up method affects ground traffic heavily and the top-down method takes longer to complete, a semi-top-down approach is often used to achieve a good balance between the two.

During a semi-top-down excavation, soils above and near the surface are excavated in accordance with the top-down procedure. Thereafter, a

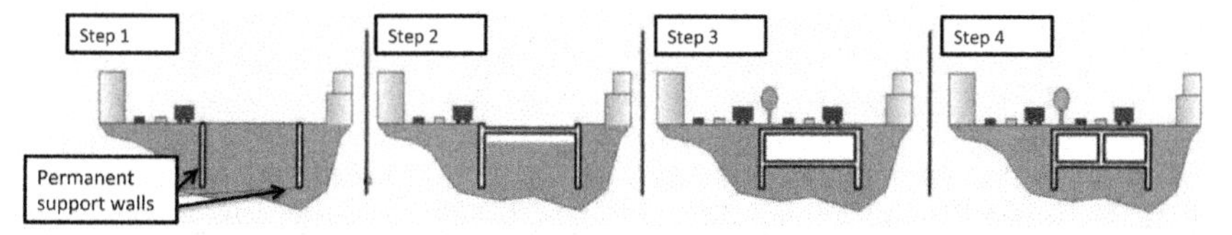

Figure 4.2 Sequence for top-down construction.

top floor slab is installed to facilitate road traffic. Deeper soils are excavated just like in the bottom-up method. Because excavation of the lower part of the pit follows a bottom-up procedure, semi-top-down excavation takes less time than is required the top-down method. Moreover, since the top floor slab is installed early in the excavation, it not only provides a stronger supporting system, but also allows traffic above to continue. Because of these advantages, semi-top-down excavation is assumed to have better performance than the other two methods in urban areas.

This method can be divided into several steps (Fig. 4.3):

1. installation of excavation support/underground structural walls, soil improvement in the trench, and dewatering if required;
2. excavation to the bottom level of top structural slabs, construction and waterproofing of top structural slabs;
3. ground surface restoration, excavation of underground interior, installation of bracing of supporting walls; and
4. construction of underground structures.

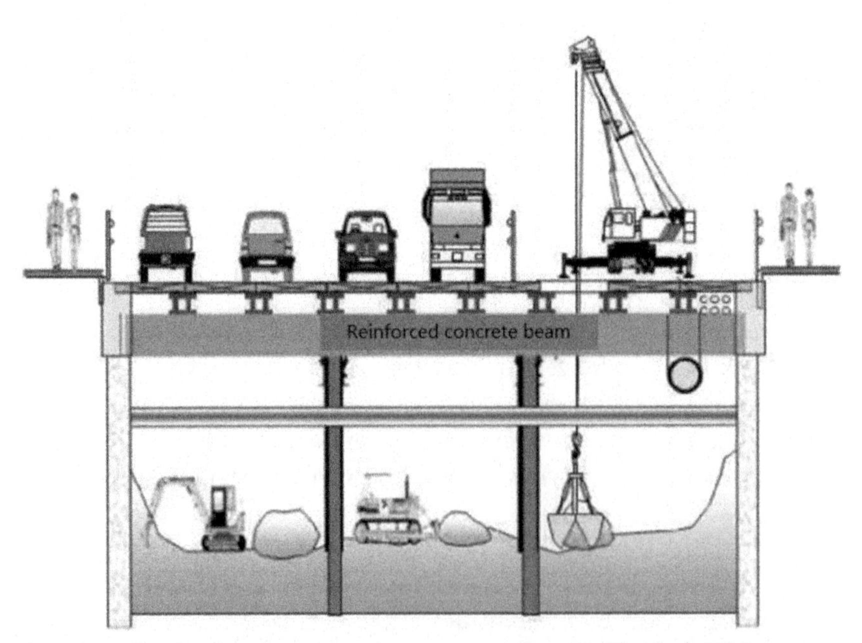

Figure 4.3 Sketch map of semi-top-down construction. *Modified from US Department of Transportation Federal Highway Administration. (2009).* Technical manual for design and construction of road tunnels—civil elements. *Publication no. FHWA-NHI-10-034.*

4.1.4 Cut-and-Cover in Soil Formation

Excavation work must be carried out in a safe working environment. This is not only to ensure the safety of workers but also to protect buildings, equipment, and other surrounding infrastructure. For soft soil, immediate support systems are needed to support the vertical or near-vertical faces of excavation and to withstand soil and water pressure. For deep excavation with very soft soil, soil improvement of pit bottom is needed before excavation.

4.1.4.1 Retaining Walls

Retaining walls (Fig. 4.4) are structures designed and constructed to resist the lateral pressure of soil when ground elevation will be changed that exceeds the angle of the repose of the soil (Ou, 2006). Retaining walls are used to support soil mass laterally so that the soil can be retained at different levels on both sides. They can bind soils between two different elevations, often in areas of terrain with undesirable slopes or in areas

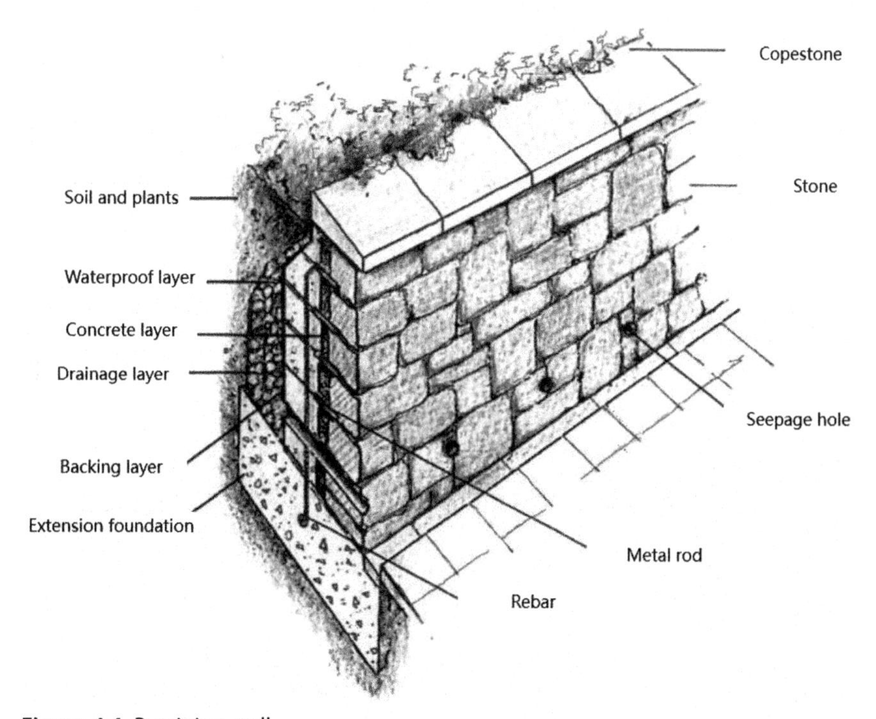

Figure 4.4 Retaining wall.

where the landscape needs to be shaped severely and engineered for more specific purposes like hillside farming or roadway overpasses.

There are four types of retaining walls (Fig. 4.5):

1. gravity walls,
2. piling walls,
3. cantilever walls, and
4. anchored walls.

1. *Steel sheet piles*

   Steel sheet piles (Fig. 4.6, left side) are widely used in civil engineering. This kind of support is applied in soft ground with a high water table. It is often used with purl-in and bracing (Fig. 4.6, right side) or even with anchor tie rods. The piles can easily be constructed and reused. But the surrounding soil is heavily disturbed and noise levels can exceed 100 dB during construction. This is why they are mostly used on rural construction sites. Large displacement and settlement may occur if grout is not injected soon after the piles are removed.

2. *Diaphragm walls*

   Also known as a slurry wall or trench, a diaphragm wall is used in deep (usually over 14 m) structures and large foundation pits due to its

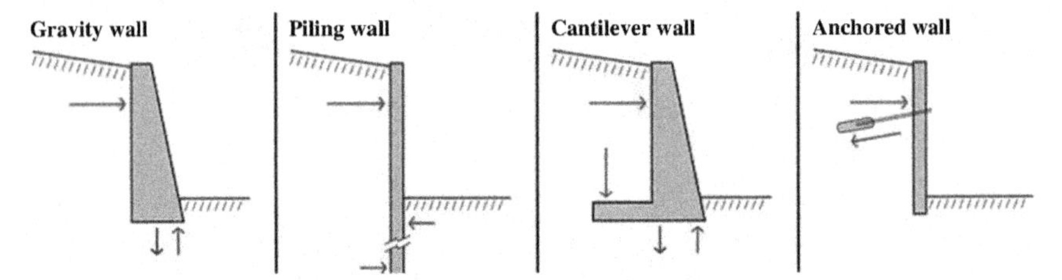

Figure 4.5 Explanation of typical retaining walls.

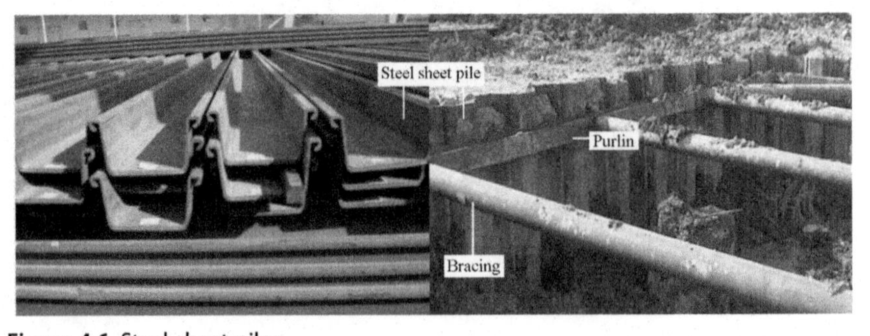

Figure 4.6 Steel sheet piles.

adaptability to almost all types of soils and its minimal impact on surroundings. Examples can easily be found in many Shanghai sites, such as at Tongji University library, Jin Mao Tower, People's Square, and the World Financial Centre.

As Fig. 4.7 shows, the construction steps are as follows:

A. preexcavation,
B. cutting of primary panel,
C. cutting of a middle section between the two primary panels,
D. installation of reinforcement,
E. concreting of primary panel,
F. cutting of secondary panel,
G. installation of reinforcement, and
H. concreting of secondary panel.

3. *Tangent bored pile walls*

A series of tangent bored piles (Fig. 4.8) can provide a relatively watertight wall. These bored piles are usually 1–3 m in diameter and extend below the bottom of the underground structure for overall stability.

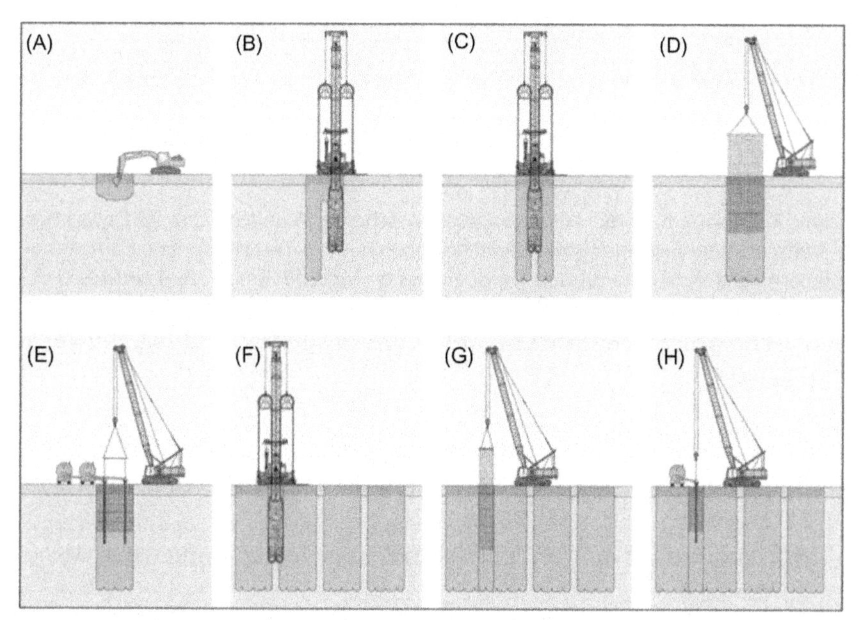

Figure 4.7 Diaphragm wall construction scheme.

Figure 4.8 Tangent bored piles.

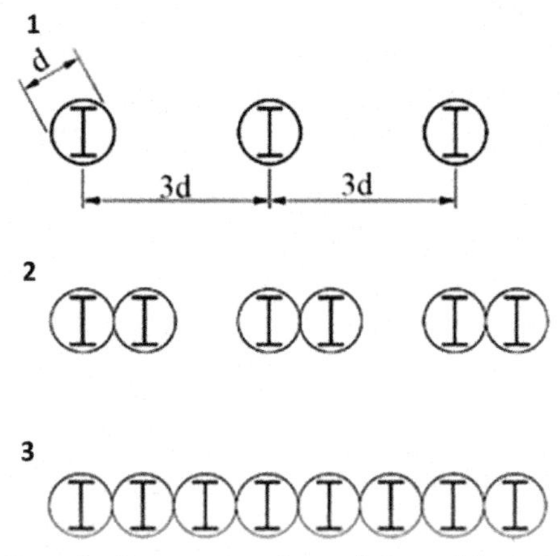

Figure 4.9 Tangent bored piles construction scheme. *Adapted from US Department of Transportation Federal Highway Administration. (2009).* Technical manual for design and construction of road tunnels—civil elements. *Publication no. FHWA-NHI-10-034.*

The general construction procedures as illustrated in Fig. 4.9 are as follows:

A. Excavation of every third bored pile. The bored piles are held open if required by the temporary casing. A steel beam or reinforcing bar cage is placed inside the shaft. The pile is then filled with concrete.

B. The next set of every third shaft is constructed in the same way as the first set.

C. The third and final set of piles is constructed, thus completing the piles.

Contrary to tangent bored piles, secant bored piles overlap each other (Fig. 4.10). For the installation of adjacent secant piles, part of the previously built bored piles is therefore removed. This pile proximity enables secant bored piles to be stiffer and more waterproof than tangent bored piles. Therefore, secant bored piles are usually used in permeable ground.

4. *Mix-in-situ piles*

Mix-in-situ piles (Fig. 4.11) are constructed using a machine that combines cement, curing agents, and soil to consolidate the ground. The piles are generally watertight, are low in cost, and have minimal environmental impact. Their use has therefore gradually become

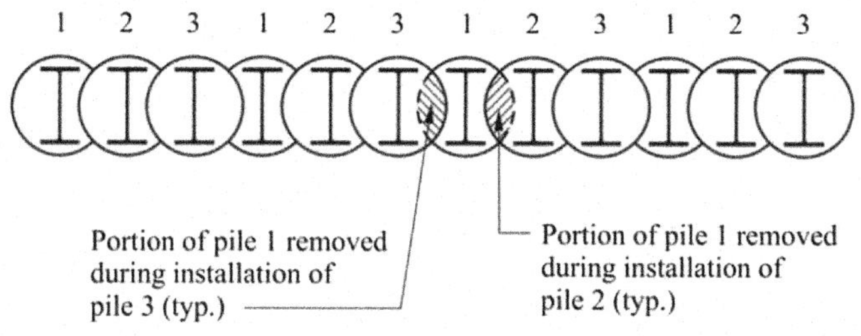

Figure 4.10 Secant bored piles construction scheme (FHWA, 2009).

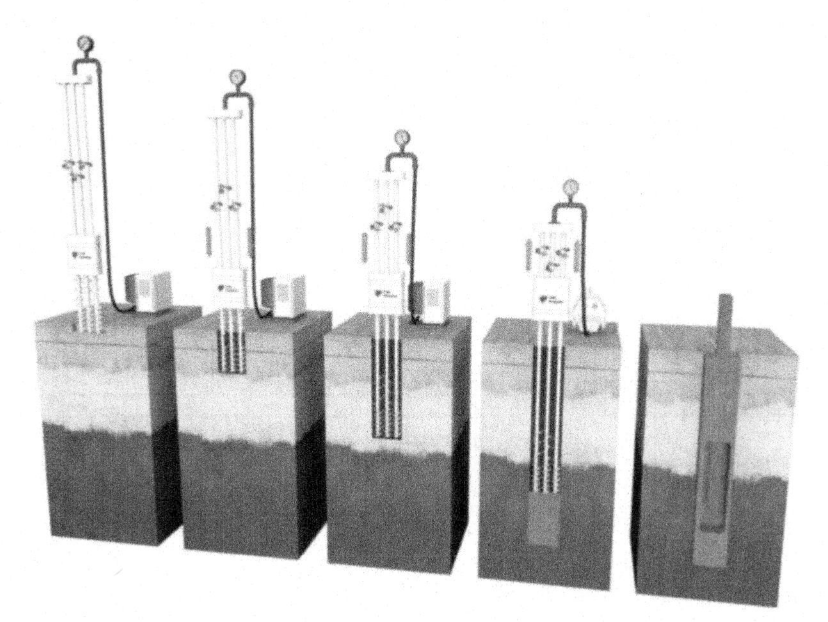

Figure 4.11 Mix-in-situ pile (Franki Foundation).

popular for excavation depths of less than 10 m in China, Japan, and other Asian countries.

The general construction procedure as shown in Fig. 4.12 is as follows:

1. setup of the auger rig at pile position,
2. stirring of the soil with the required size auger,
3. pumping of cement grout while stirring,
4. double mixing by auger rotation as a form of mechanical agitation,
5. withdrawal of the auger with counterclockwise rotation, and
6. installation of grout-mix column.

5. *Soil-mixing walls*

   Contrary to mix-in-situ piles, soil-mixing walls use strengthened elements such as steel H-piles (Fig. 4.13) that are inserted into the completed soil-mixing wall (Fig. 4.14). This method is a cost-effective technique for the construction of walls (typically 6–15 m) for groundwater control, excavation support, and ground improvement. The main benefits are high efficiency, low water permeability, limited spoil, and minimal vibrations during construction.

6. *Soil nail walls*

   Soil nail walls are used to strengthen a trench by inserting steel reinforcement bars into the soil and anchoring them to the soil strata. Soil nailing is frequently used in soft areas where landslides might occur. It is not recommended to use in soft clay soils, silts, and sands where inner soil cohesion is low, as well as in areas of below water table.

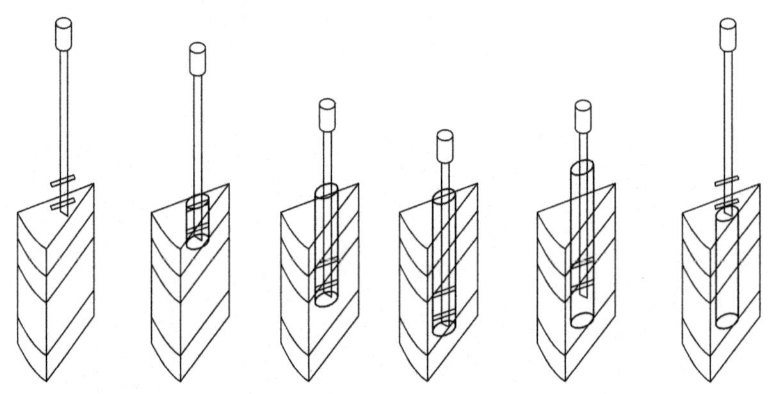

Figure 4.12 Mix-in-situ pile construction scheme.

Figure 4.13 Reinforcing elements.

Figure 4.14 Mixing machine (Bauer, 2012).

The installation process as illustrated in Fig. 4.15 is as follows:

1. excavation of an initial lift;
2. drilling of nail hole;

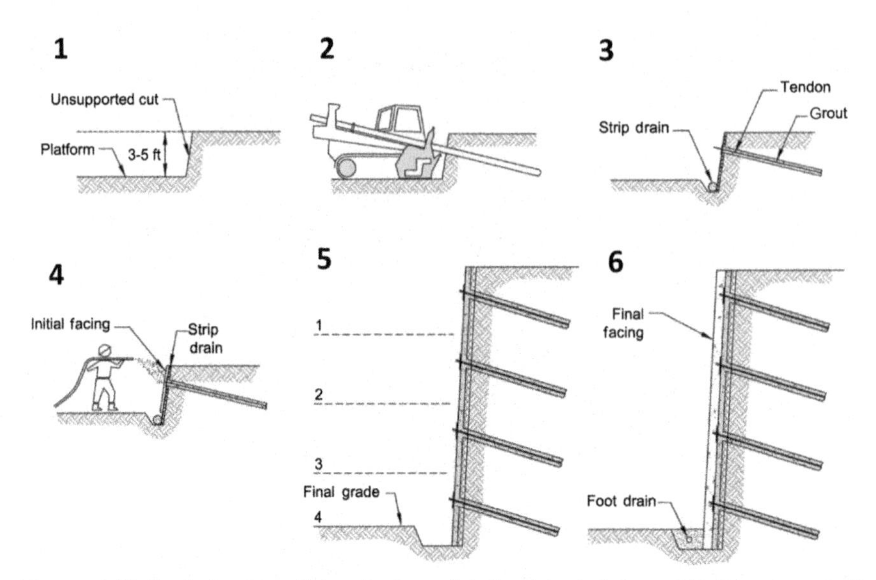

Figure 4.15 Typical soil nail head plate details. *Adapted from US Department of Transportation Federal Highway Administration. (2009).* Technical manual for design and construction of road tunnels—civil elements. *Publication no. FHWA-NHI-10-034.*

3. installation and grouting of nail that includes installation of a strip drain;
4. placement of initial facing by shotcrete, reinforcement, bearing plate, washer and hex nut installation;
5. construction of subsequent levels; and
6. placement of final facing including building of a foot drain.

4.1.4.2 Pit Excavation and Support

Pit excavation involves complicated works such as the need of trenchs to protect the excavation from groundwater ingress and sometimes the difficult surrounding conditions which impose restructions during construction. The size of pit is one of the factors that define the construction process, however a sequence of works to be done are described below.

1. *Dewatering*

   In ground with high water, dewatering is performed by burying a series of well tubes around the bottom of the pit before excavation (Fig. 4.16).
2. *Earth excavation and lateral supports*
   a. Trench digging into the first lateral support level
   b. Lateral supporting

c. Continuing excavation into the second support level
d. Lateral supporting
e. Looping excavation and support until pit bottom level is completed

4.1.4.3 Structure Construction and Backfilling

An underground structure is usually built using reinforced concrete. Reinforced concrete engineering can be divided into three steps:

1. *Template*

As a model for new casting of concrete, the formwork is a necessary construction material and tool in civil engineering. The quality of the manufacture and installation of the template will directly affect the quality of the concrete structure, and its functionality; its ease of disassembly and installation as well as its reusability will also affect the construction speed and the cost of the concrete engineering. Therefore, according to the structural and construction conditions of the project, the proper template form, template structure, and construction method should be chosen in order to ensure the quality and safety of the concrete construction and to minimize time and cost.

Template materials are made of wood, plywood, steel, aluminum alloy, plastic, fiber-reinforced plastic, precast concrete sheet, profiled steel plate, and so on. The template structure consists of a panel, a skeleton, a support system, an operation platform, steel rings, and connectors (Fig. 4.17).

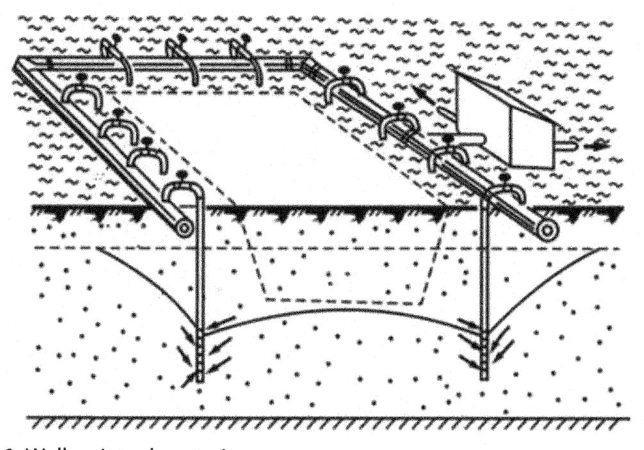

Figure 4.16 Well-points dewatering.

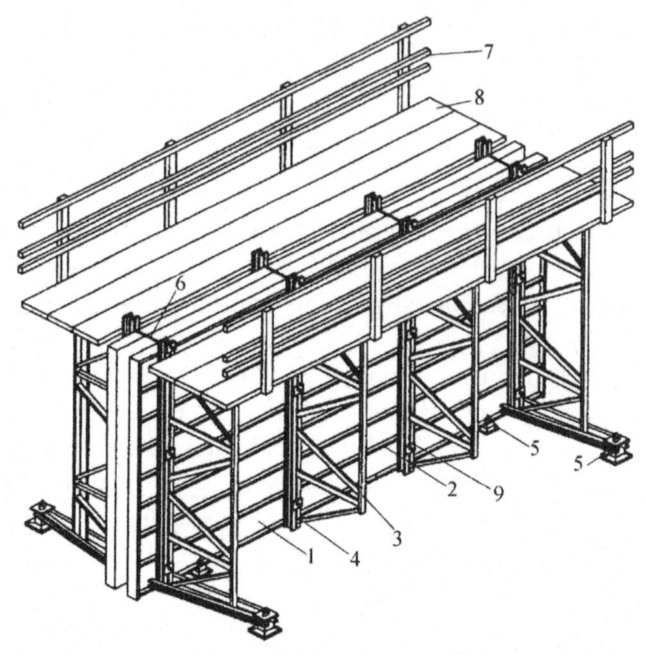

Figure 4.17 Template construction (1. Panel; 2,3,4. Skeleton; 5. Support system; 6. Connectors; 7. Guard rail; 8. Operation platform; 9. Steel rings).

2. *Steel bar*

   A steel bar is a very important part of reinforced concrete engineering. The correct placement of its position, quantity, and specification directly affects the ability of the structure to bear external force.

   The main method of mechanical connection of steel bars in a foundation pit is threaded connection. The threaded steel sleeve connection includes conical thread connection and straight-thread connection. Using the principle that end connection threads can withstand axial and horizontal force, moreover it has good self-locking properties as steel bars are connected by specified mechanical force as is shown in Fig. 4.18.

3. *Concrete*

   The pouring of concrete needs to prevent any rebar displacement and ensure the uniformity, compactness, and integrity of the concrete mixture. Because of the characteristics of underground structures, it is impossible to complete the irrigation all at once. Therefore, construction joints should be set up, usually in places where the tensile force and the shear force are minimum, also taking into consideration the convenience of the construction work.

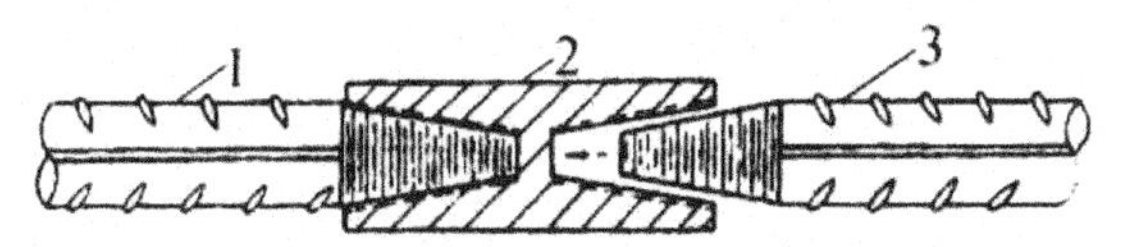

Figure 4.18 Connection of rebar taper threaded sleeve (1. Connected steel bar; 2. Conical-threaded sleeve; 3. Unconnected rebar).

In order to ensure that the difference between the internal and external concrete temperature is less than 25°C, we need to measure the concrete temperature regularly both internally and externally. Based on the results, corresponding measures are carried out to avoid or reduce cracks in mass concrete.

The working procedure after the completion of the structure is backfilling and then restoring of the original ground.

4.1.5 Cut-and-Cover in Rock Formation

According to uniaxial compressive strength, rocks can be classified into six grades: extremely hard, very hard, hard, intermediate hard, weak, very weak, and extremely weak. Within the depth of the cut-and-cover method, there are usually medium hard, weak, and very weak rocks. Support systems are therefore often necessary to guarantee safety.

Some of the methods mentioned for soft soils, such as soil nail walls, can also be applied in weak rock conditions.

4.1.5.1 Retaining Wall Construction

1. *Anchored retaining walls*

   An anchored retaining wall is comprised of a reinforced concrete wall and anchors that rely on the horizontal tension of the rock bolt to bear the lateral soil pressure. This method is suitable for rock slopes and areas that are hard to excavate. Anchored retaining walls are either the panel type or column type (Fig. 4.19). The construction process is as follows:

   1. cleaning of the slope and drilling of holes,
   2. insertion of anchor bolts into the holes and grouting,
   3. installation of anchor head, and
   4. construction of retaining wall.

2. *Micropiles*

   Micropiles (Fig. 4.20) generally refer to bored piles with a diameter of under 300 mm. The method has been widely employed due to

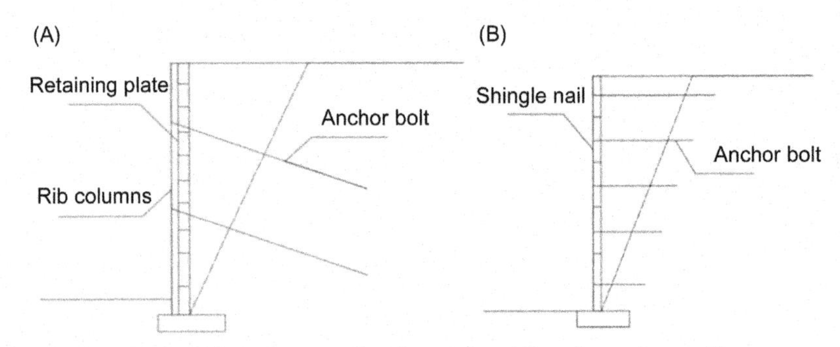

Figure 4.19 Anchored retaining wall: column type (A) and panel type (B).

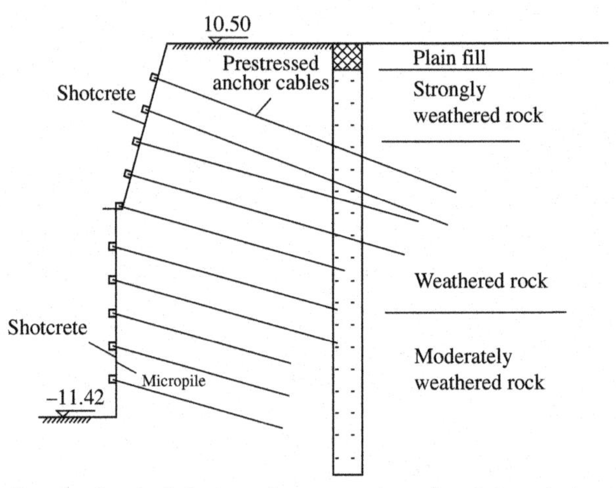

Figure 4.20 Sketch of typical design of cross section of rock foundation pit.

its convenience and simplicity in construction. It is also favorable in limited areas because of its minimal vibration and impact on surroundings.

In Qingdao, micropiles are used together with rock bolts. This can effectively decrease the horizontal displacements of the foundation pit and redistribute the tension in the prestressed anchor cables.

3. *Pit excavation*

A pit in rock formation is more stable than in soil formation. It is therefore not necessary to strengthen the ground as in soils. Bucket excavation is not possible in rock formation as the excavation media is too hard. The drill-and-blast method or hydraulic hammer excavation are usually applied in rock formation.

4.2 CONVENTIONAL TUNNELING

The conventional tunneling method is a cyclical process of tunnel construction that involves excavation by drilling and blasting or by mechanical excavators (except the full-face tunnel boring machine (TBM)). This is followed by application of an appropriate primary support.

Mainly used in rock formations, several variants exist including the mining method, drill-and-blast method, new Austrian tunneling method, etc. Conventional tunneling has the following steps (Fig. 4.21):

1. drilling of blast holes,
2. charging,
3. ignition,
4. ventilation,
5. mucking: loading and hauling,
6. scaling,
7. temporary support, and
8. surveying

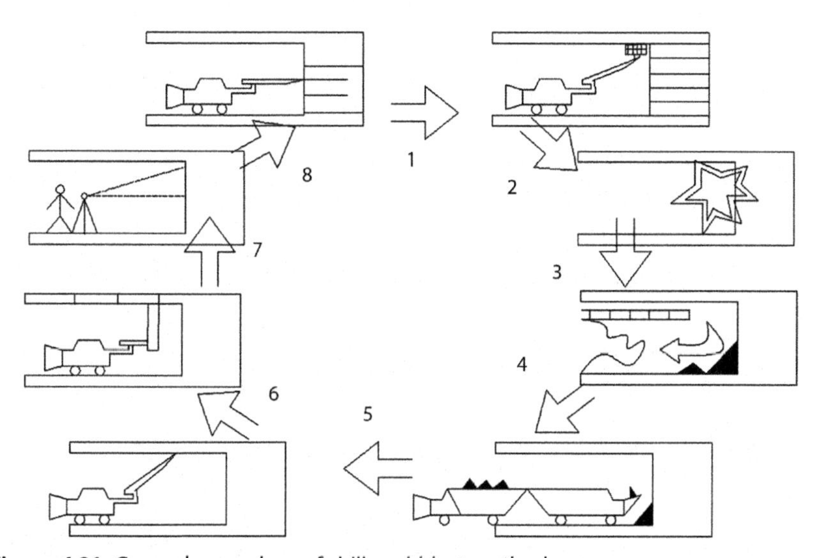

Figure 4.21 General procedure of drill-and-blast method.

1. *Drilling of blast holes*

   Before blasting it is necessary to drill holes in the rock to insert explosives. These holes are drilled in a pattern (Fig. 4.22) by drill rigs. The three types are:

   1. Parallel cuts. They are usually used in hard and intact rocks. The direction should be precisely controlled.
   2. V cuts. They are applicable to all kinds of rock and no large-diameter drilling machine is needed. The drilling angle is hard to control.
   3. Fan cuts. They are not as common as the two previous cuts due to the asymmetry of blast holes and delicately designed position and depth. This cut is suitable for rocks with fractures.

2. *Charging*

   Once the pattern has been drilled, the depth of its holes is verified and the presence of water for efficiency and safety checked. The holes are then charged with a detonator and a primer is lowered to the bottom of the hole. Explosives are then pumped down the hole around the detonator and primer, before the end of the holes are filled with stemming. This acts as a plug and forces the explosive energy to go into the surrounding rock rather than along and out of the hole. Once all the holes have been charged, they are connected to explode in a certain order.

   Regarding the choice of explosives, several variations exist:

   - Ammonium nitrate/fuel oil. Cheap and bulk dosed but sensitive to water.
   - Emulsions. Pumpable, resistant to water, and thus growing in use.
   - Explosive gelatines that are mainly used for smooth blasting.
   - Explosive powders that are less and less commonly used.

   It is also possible to use expansive cements, which are soundless chemical demolition agents that do not produce an explosion. But they take more than 10 hours to work.

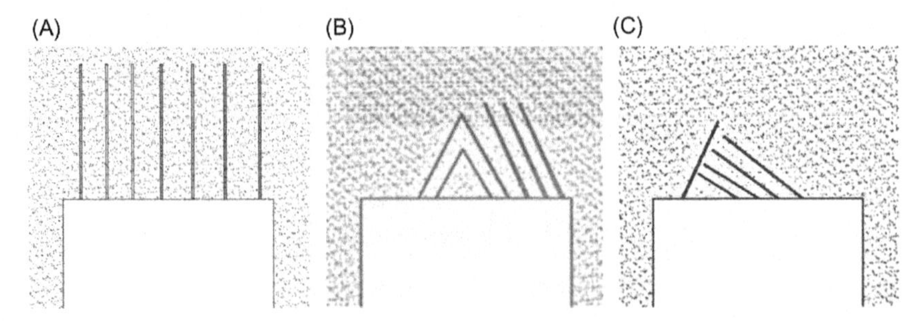

Figure 4.22 Blast hole patterns: (A) parallel cut; (B) V cut; and (C) fan cut.

3. *Ignition*

   The ignition system can include different elements.
   - Electric detonators (Fig. 4.23) used to initiate explosives such as blasting works in noncoal mines, open pits, and demolition and other engineering projects.
   - Nonelectric detonators. They have the advantage of enabling fast and easy connection of all detonators at one ignition point (compared to the time-consuming and complicated process of connecting electric detonators).
   - Detonating cords. They are used to reliably and cheaply chain together multiple explosive charges. Typical uses include mining, drilling, demolitions, and warfare.
4. *Ventilation*

   Ventilation is necessary to secure a safe and healthy working environment in the tunnel (Fig. 4.24) to exhaust smoke, poisonous gas, and dust that are produced while drilling and blasting.
5. *Scaling*

   Scaling is the removal of insecure blocks of rock from the back, the sidewalls, and the face. It is also regarded as preparation for shotcrete and/or rock bolts. Due to its danger, the safety of workers must be carefully considered. Procedures vary throughout the world, from manual to completely mechanized (Fig. 4.25).
6. *Mucking: loading and hauling*

   To make enough room to operate and ensure a good environment for workers in the tunnel, muck should be removed. Thus, after scaling muck is loaded (Fig. 4.26).

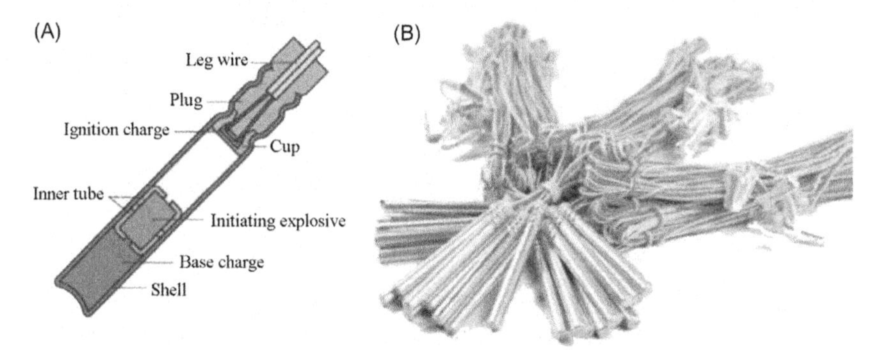

Figure 4.23 Electric detonators: image of electric detonators for rocks.

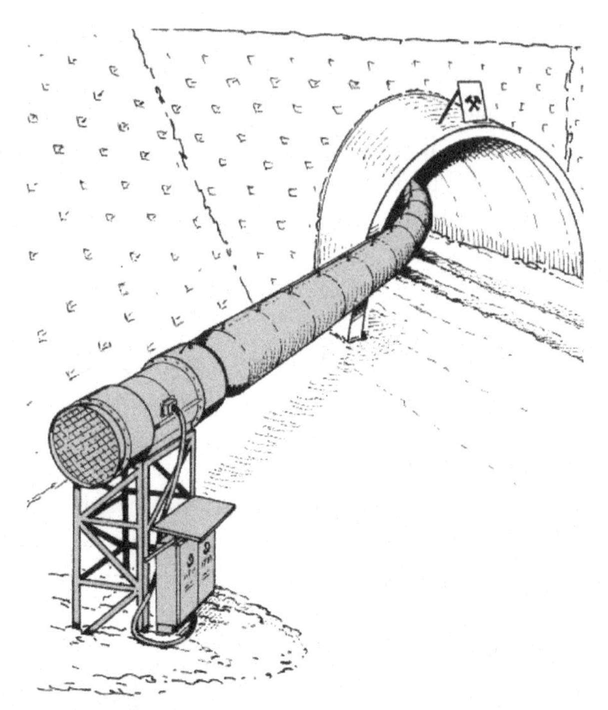

Figure 4.24 Tunnel ventilation.

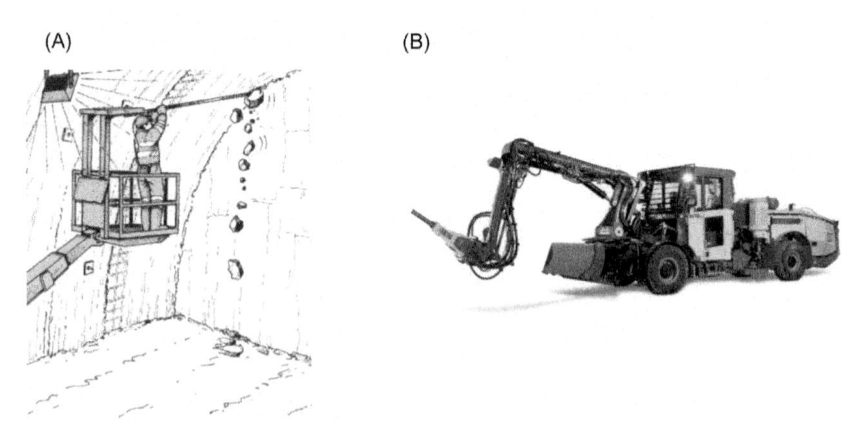

Figure 4.25 Scaling: (A) manual scaling and (B) mechanized scaler (Mining & Construction, 2013).

While hauling, niches or bays are needed for trucks to turn around when the tunnel is small (Fig. 4.27). The number and distance between them is decided such that no delay in construction is caused.

Figure 4.26 A mucking loader.

Figure 4.27 Hauling mucks.

7. *Temporary support*

After excavation, temporary support must be provided to prevent rock fall or collapse. The different forms of support will be introduced and developed in Section 4.2.2.

8. *Surveying*

A geologist maps the excavation face, to be able to adapt the excavation process according to the geology that is encountered. Rock behavior and water infiltration are also noted.

4.2.1 Excavation Method

Similar to the support method, the excavation method depends on the ground conditions. Conventional tunneling allows either full-face or partial excavation of the tunnel cross section. An appropriate excavation sequence must be defined on the basis of the expected geotechnical conditions, tunnel size, modeling results, structural analysis, and experience.

As an example, the application conditions of the excavation methods are introduced here, according to the Chinese code rock classes (Table 4.1).

4.2.1.1 *Full-Face Excavation*

Full-face excavation is normally used for small-sized tunnels of rock class I to III. The excavation tool can be a jumbo (Fig. 4.28) or handheld airleg for a small-scale excavation when access is difficult.

Table 4.1 Rock mass classes according to Chinese design codes (The National Standards Compilation Group of People's Republic of China, 1994)

Rock mass class	Rock mass basic quality classification index, BQ	Velocity of longitudinal elastic wave in the rock mass, V_{pm} (km/s)
I	>550	>4.5
II	550–451	3.5–4.5
III	450–351	2.5–4.0
IV	350–251	1.5–3.0
V	≤ 250	1.0–2.0
VI		<1.0 (for saturated soil <1.5)

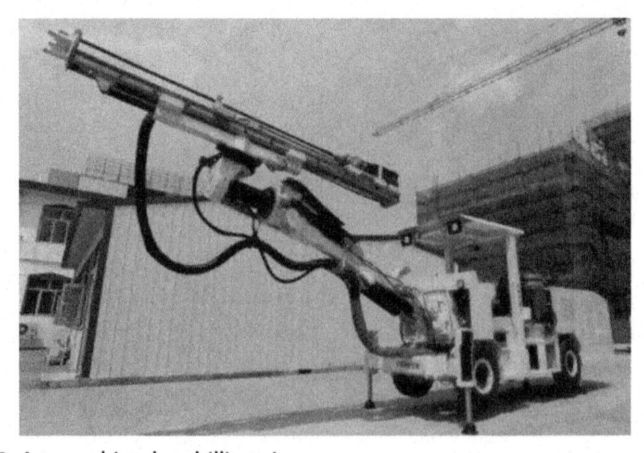

Figure 4.28 A tunnel jumbo drilling rig.

4.2.1.2 Stage Excavation

The stage excavation is generally used for classes IV and V. Depending on the dimensions, stage excavation can be classified into three categories (K is defined in Fig. 4.29):

1. A small section of long-stage excavation ($K > 50$ m) is needed to maintain the stability of the excavation face.
2. A short-stage method ($K > 5-50$ m) can improve the stress conditions of initial supports to control the deformation of surrounding rocks. Nevertheless, muck in the upper stage may influence the construction of the lower stage.
3. When the stage is 3–5 m long ($K = 3-5$ m), it is known as an ultra-short stage method. But the interference between the upper and lower stage is significant, which means the excavation efficiency decreases.

4.2.1.3 Crown Heading Excavation

For crown heading excavation, the bench and invert excavation is only started after the excavation of the tunnel crown (Fig. 4.30). This method is typically useful when the area of excavation is in fragmentary strata. Crown heading is normally used for classes IV or V and if the tunneling size is larger, classes III or IV.

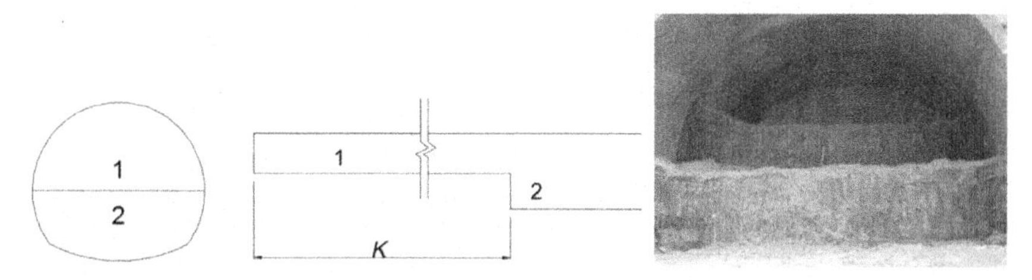

Figure 4.29 Stage excavation.

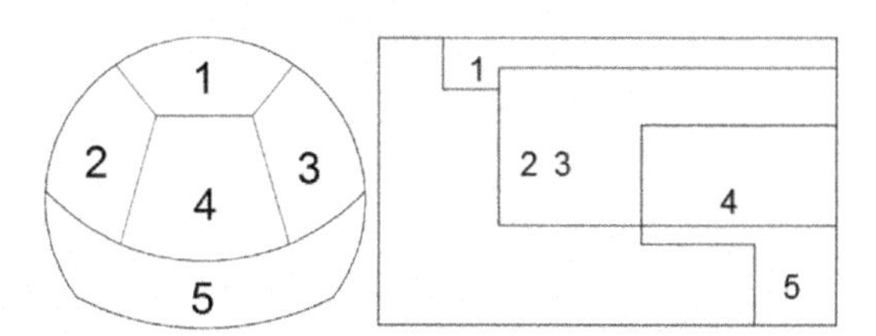

Figure 4.30 Crown heading excavation.

4.2.1.4 CD Excavation

CD excavation (midwall method, Fig. 4.31) is normally used for classes IV and V whose span is less than 18 m. There is a greater chance of settlement and the middle temporary support has to be removed later.

4.2.1.5 CRD Excavation

Center cross diaphragm method (CRD) excavation (cross-midwall method, Fig. 4.32) is used for fractured rocks as it provides top and lateral supports. It has been adopted in many projects, but the construction period is long and the cost for temporary support construction and removal is high. It is normally used for large caverns for rock class III and for smaller ones, classes IV and V.

4.2.1.6 Sidewall Adit Heading Excavation

The sidewall adit heading method (Fig. 4.33) is advantageous in weak rock (class V) such as clay and silt and for tunnels with shallow depth and

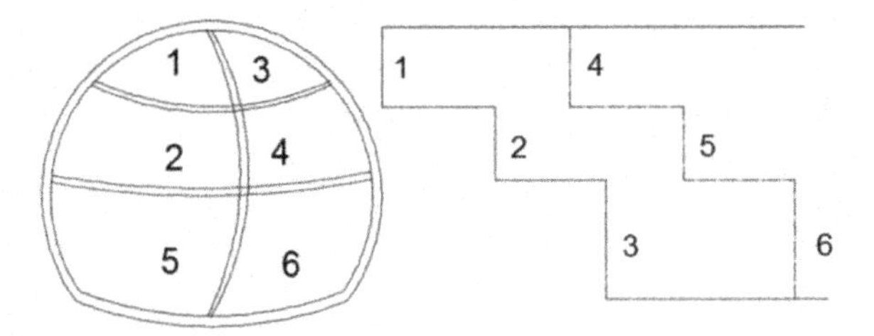

Figure 4.31 CD excavation.

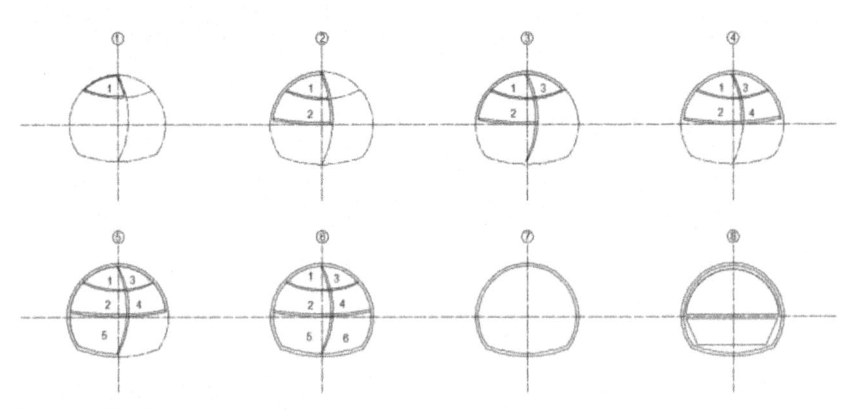

Figure 4.32 CRD excavation.

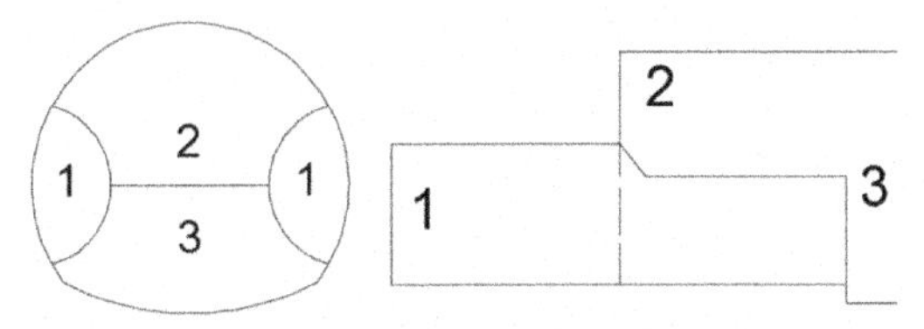

Figure 4.33 Sidewall adit heading excavation.

Figure 4.34 Xiangan tunnel under construction.

wide span. Its drawbacks are lengthier construction periods, higher costs, and weaker waterproofing. The Xiangan tunnel in Xiamen (Fig. 4.34) uses this construction method.

4.2.2 Support System

4.2.2.1 Anchors and Rock Bolts

Anchors and rock bolts (Fig. 4.35) involve the same mechanical principle, but they are employed in soft soil and rock, respectively. Anchors and rock bolts are active reinforcing elements designed to anchor and stabilize the rock mass during tunnel excavations. In case of rock mass movement, the bolts will develop a restraining force that is transferred back to the rock mass as in Fig. 4.36. The driving force is countered in this way, such that the total resistance mobilized within the rock mass is equal to the applied force.

Figure 4.35 Rock bolt installation (Dywidag-Systems International, 2014).

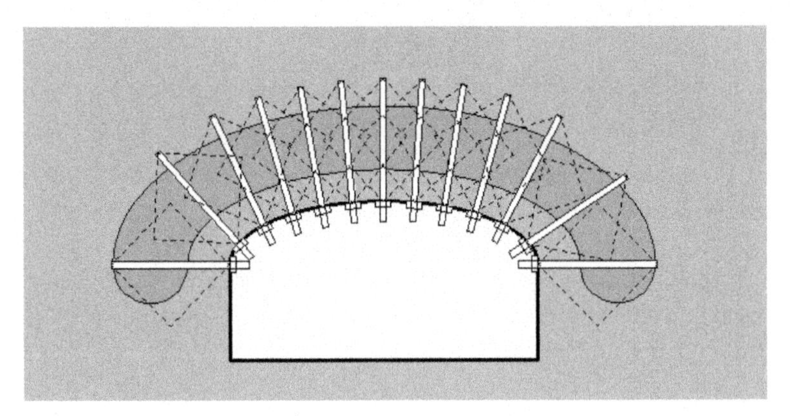

Figure 4.36 Support system anchors.

4.2.2.2 Lattice Girders

Lattice girders (Fig. 4.37) are lightweight triangular steel frames. In crown heading excavation, for example, immediate support should be applied to the excavation with lattice girders. They are used in conjunction with shotcrete and act as armor to achieve a good level of support in tunneling. To provide additional forward support, strata bolts can be inserted through the lattice girders (Tunnel Ausbau Technik, n.d.).

Figure 4.37 Tunnel lattice girders.

Figure 4.38 I beam girders.

I beam girders (Fig. 4.38) are relative heavy steel frames. They are widely used in China tunneling projects for weak rock formation. However, the use of I beam girders will have a bad influence on shotcrete quality due to their geometric shape.

4.2.2.3 Sprayed Concrete

Sprayed concrete is a fast hardening material used to stabilize and support tunnels. It comes in two forms: gunite (dry mix) and shotcrete (wet mix). Shotcrete is more efficient than gunite as shotcrete has a lower rebound rate.

Dry sprayed concrete is used when small amounts and high early strength is needed (Höfler, Schlumpf, & Jahn, 2011). The material is conveyed in a dry or semidry state to the nozzle, where water is added to the mix before being applied at high velocity onto the substrate.

Shotcrete is used more when a high concrete quality is required (Höfler et al., 2011). It requires a mechanized nozzle, as the wet mixture is too heavy to be held by a worker (Fig. 4.39). This wet sprayable concrete is workable, premixed, and consists of aggregate, cement, and water. For spraying, wet-sprayed concrete is mixed with air and shotcrete accelerators before being applied. Fibers are frequently applied in shotcrete to achieve better performance in terms of tensile strength, durability, and antifire function of the lining.

4.2.2.4 Presupport

Presupport measures in fractured, yet good rock mass aim to increase standup time. This limits the overbreak, ensuring safe excavation and

Figure 4.39 Sprayed concrete system.

Figure 4.40 Spiles.

allowing efficient initial support installation. Spilling (Fig. 4.40) and forepoling are two common presupport measures.

The former is the insertion of tensile strengthening elements in the ground such as dowels. The latter involves the installation of steel sheets or bars. In cases where the rock mass is severely fractured, self-drilling rock reinforcement pipes are used to avoid borehole collapse (U.S. Department of Transportation Federal Highway Administration (FHWA), 2009).

4.2.3 Waterjet Technology

Blasting technologies cause serious ground vibrations that can be reduced by wave-screening methods. Waterjet precutting is one recent example technology. It involves an abrasive waterjet that performs precise precutting to prevent the propagation of elastic waves (Oh, Joo, Hong, Cho, & Ji, 2013).

4.3 MECHANICAL TUNNELING

TBMs are extensively used throughout the world to bore various tunnel diameters in various types of rock and soil strata. A TBM consists of a main body and supporting elements that enable it to perform cutting, shoving, steering, gripping, shielding, exploratory drilling, ground control and support, lining erection, muck removal, and ventilation (FHWA, 2009).

4.3.1 Tunnel Boring Machine Types

There are several types of TBMs (Fig. 4.41). The best TBM for a project is based on the geological conditions of the site and the project's features. The general classification of the different types of TBMs for both hard rock and soft ground are presented here.

4.3.1.1 Gripper Tunnel Boring Machine

A gripper TBM (Fig. 4.42) is suitable for driving in hard rock conditions when there is no need for a final lining. The rock supports (rock anchors,

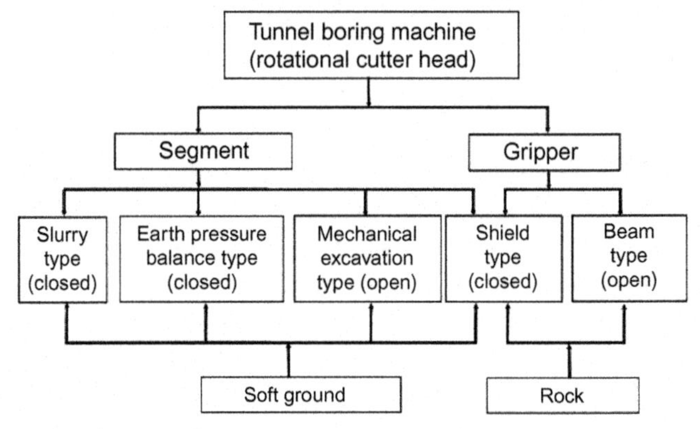

Figure 4.41 Classification of TBMs.

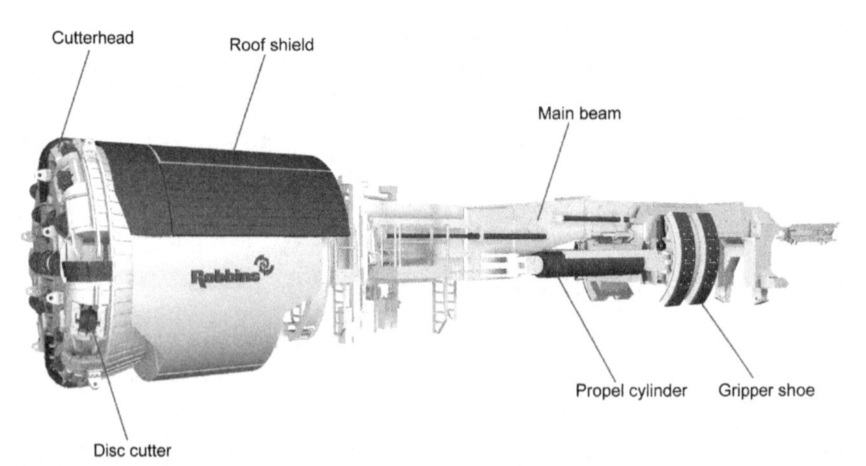

Figure 4.42 Typical diagram of an open gripper main beam TBM. *Courtesy of The Robbins Company.*

wire-mesh, shotcrete, and/or steel arches) can be installed directly behind the cutter head shield and enable controlled relief of stress and deformations. The existence of mobile partial shields enables gripper TBMs to be flexible even in high-pressure rock. This is useful when excavating in expanding rock to prevent the machine from jamming.

4.3.1.2 Double-Shield Tunnel Boring Machine

A double-shield TBM is generally considered to be the fastest machine for hard rock tunnels under favorable geological conditions with installation of the segment lining. It is possible to drive 100 m in 1 day. This type of TBM consists of a rotating cutter head and double shields (Fig. 4.43), a telescopic shield (an inner shield that slides within the larger outer shield), and a gripper shield together with a shield tail.

Figure 4.43 Typical diagram of a double-shield TBM. *Courtesy of The Robbins Company.*

While boring, gripper shoes radially press against the surrounding rock to hold the machine in place and take some of the load from the thrust cylinders. For the motion of the front shield the gripper shoes are loosened, before the front shield is pushed forward by thrust cylinders protected by the extension of the telescopic shield. Because regripping is a fast process, double-shield TBMs can almost continuously drill. As for the shield tail, it is used to provide protection for workers while erecting, installing the segment lining, and pea grouting.

4.3.1.3 Single-Shield Tunnel Boring Machine

Single-shield TBMs (Fig. 4.44) are used in soils that do not bear groundwater and where rock conditions are less favorable than for double shields, such as in weak fault zones. The shield is usually short so that a small radius of curvature can be achieved.

4.3.1.4 Earth Pressure Balance Machines

Earth pressure balance (EPB) technology (Fig. 4.45) is suitable for digging tunnels in unstable ground such as clay, silt, sand, or gravel. An earth paste face formed by the excavated soil and other additives supports the tunnel face. Injections containing additives improve the soil consistency, reduce soil stick, and thus its workability.

To ensure support pressure transmission to the soil, the earth paste is pressurized through the thrust force transfer into the bulkhead (Facesupport.

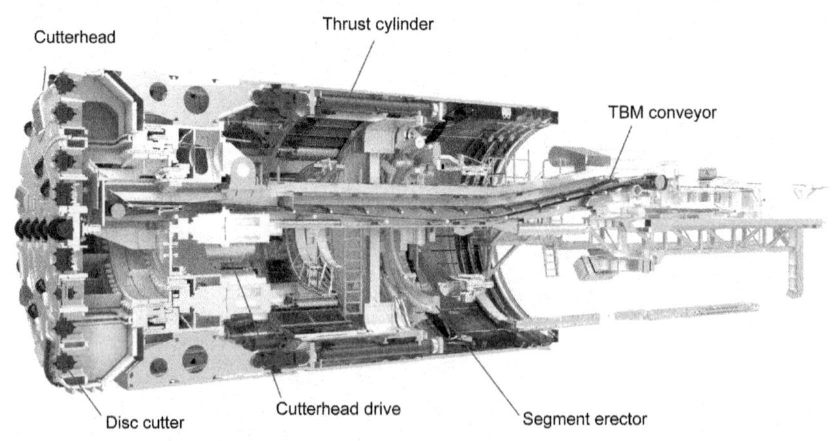

Figure 4.44 Typical diagram of single-shield TBM. *Courtesy of The Robbins Company.*

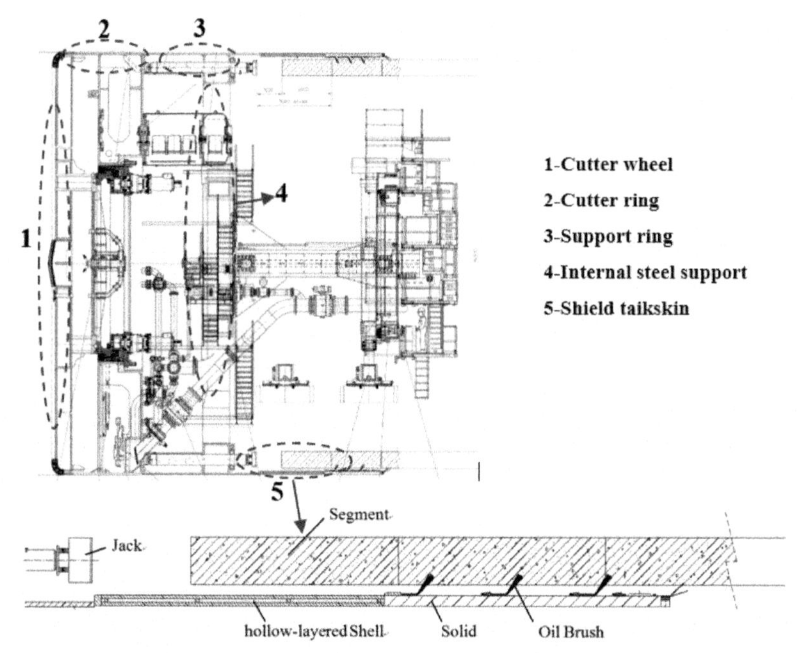

Figure 4.45 Scheme of an EPB machine.

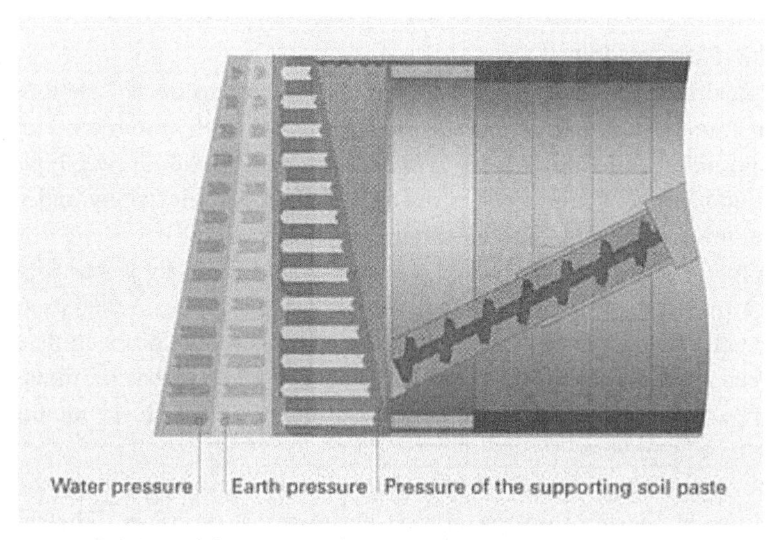

Figure 4.46 Scheme of the pressure for an EPB. *Courtesy of Herrenknecht AG.*

org (FS), n.d.). The TBM advance rate (inflow of excavated soil) and the soil outflow from the screw conveyor regulate the support pressure at the tunnel face. This is monitored at the bulkhead by the readings of pressure sensors (Fig. 4.46).

4.3.1.5 Slurry Tunnel Boring Machine

Slurry TBMs are used for highly unstable and sandy soil and when the tunnel passes beneath structures that are sensitive to ground disturbances (NFM Technologies, n.d.). Pressurized slurry (mostly bentonite) supports the tunnel face. The support pressure is regulated by the suspension inflow and outflow. The slurry's rheology must be chosen in accordance with the soil parameters and should be carefully and regularly monitored (FS, n.d.).

4.3.1.6 Mixshield Technology

Mixshield technology (Fig. 4.47) is a variant of conventional slurry technology for heterogeneous geologies and high water pressure. In mixshield technology an automatically controlled air cushion controls the support pressure, with a submerged wall that divides the excavation chamber. This wall seals off the machine against the excess pressure from the tunnel's face. As air is compressible in nature, the mixshield is more sensitive in pressure control and thus will provide more accurate control of ground settlement.

4.3.1.7 Pipe Jacking

Pipe jacking (also called microtunneling) is a micro- to small-scale tunneling method for installing underground pipelines with minimum surface disruption. It is used for sewage and drainage construction, sewer replacement and lining, gas and water mains, oil pipelines, electricity and telecommunications cables, and culverts (PJA, 2017).

A fully automated mechanized tunneling shield is usually jacked forward from a launch shaft toward a reception shaft. Jacking pipes are then progressively inserted into the working shaft. Another significant difference between the pipe jacking method and shield method is that the lining of the pipe jacking is made of tubes and the lining of the shield method is made up of segments.

In order to significantly reduce the resistance of the pipes, a thixotropic slurry is injected into the outside perimeter of the pipes. The thixotropic slurries can also reduce disturbance to the ground while pipe jacking slurry thickness. The thickness should be six to seven times the void between the machine and pipes.

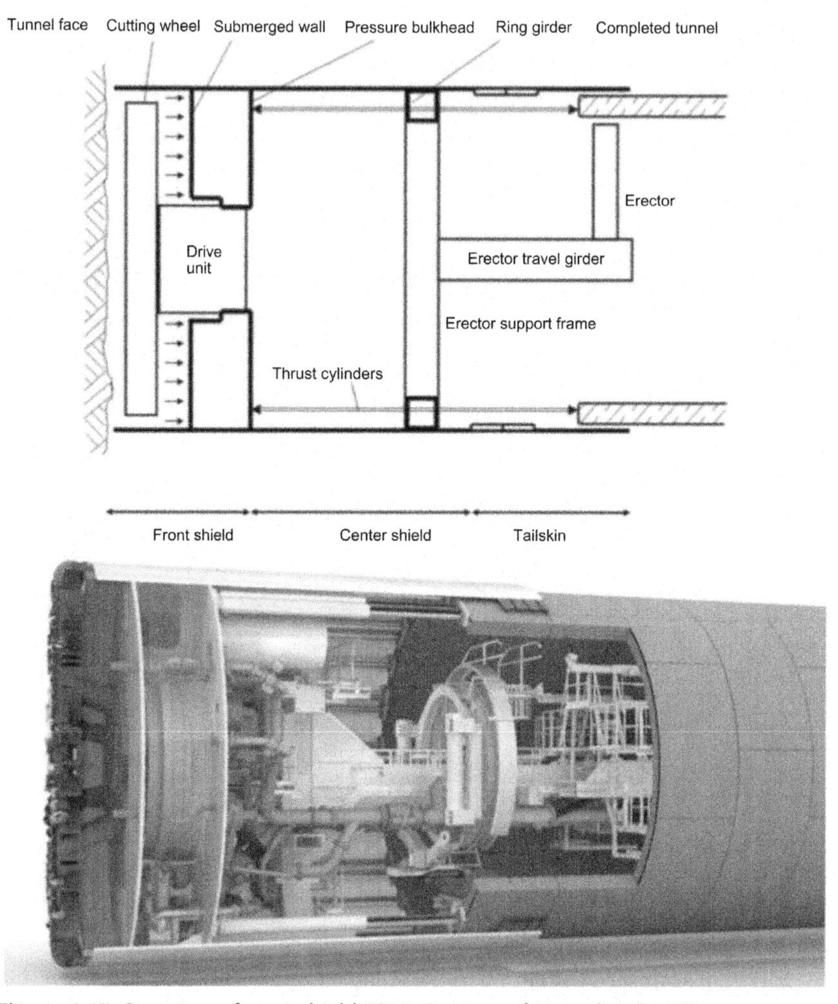

Figure 4.47 Overview of a mixshield TBM. *Courtesy of Herrenknecht AG.*

4.3.1.8 Partial-Face Excavation Machine

Partial-face excavation machines (Fig. 4.48) have an open-face shield and can sometimes be more economical in homogeneous and semistable ground with little or no groundwater (Herrenknecht, n.d.b). In boulders layer, the open-face can deal with boulders much easier than closed shield machines. In cavity ground, the open-face can avoid the risk of falling down into the bottom of the cavity. Thanks to their simple design and that the operator workplace is close to the open tunnel face,

Figure 4.48 Two kinds of partial-face excavation machines. *Courtesy of Herrenknecht AG.*

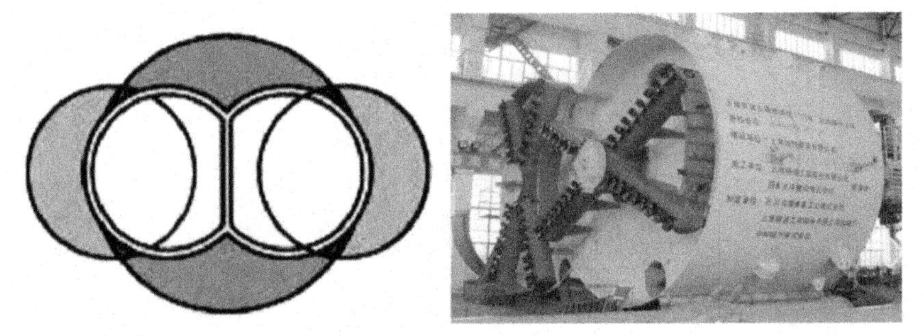

Figure 4.49 Cross section of DOT shield.

these machines can easily be adapted to changing geological conditions. Good excavation monitoring can also be carried out.

4.3.1.9 Noncircular Shields

Only approximately two-thirds of a circular tunnel section can effectively be used. Consequently, the TBMs of the future are expected to have noncircular cross-sectional machines and be so-called noncircular shields. Different types already exist, such as double-O-tube shield tunneling (DOT) shields (Fig. 4.49) for which a middle column is installed in order to form a stable tunnel lining.

Various tunnel cross sections (Fig. 4.50) can also be made using a shield machine with a primary circular disk cutter in the center and multiple secondary planetary cutters on the peripheries (Fig. 4.51).

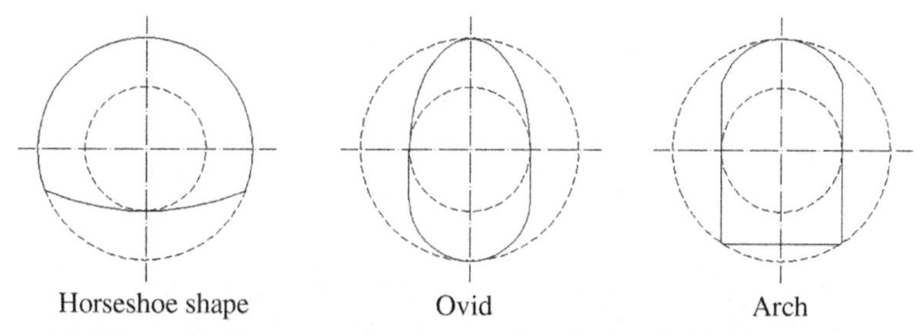

Figure 4.50 Example of cross section that can be obtained with TBMs.

Figure 4.51 A rectangular shield.

In some projects in China, chips with fabrication information are imbedded inside segments.

4.3.1.10 Deep Drilling Technology

Today, different companies have been working on developing technology that can cope with specific demands. Oil, gas, and geothermal energy sources can be explored using deep drilling rings. The Terra Invader type deep drilling rig is an efficient drilling technology used to explore deep energy deposits and can be used for onshore and offshore drills to depths of down to 8000 m (Fig. 4.52).

4.3.2 Excavation Tools

The wear of excavation tools considerably affects the economic and tunneling performance of a TBM. Different excavation tools exist for different ground conditions. Examples include cutting bits, cutting knives, disk cutters (Fig. 4.53), and rippers. Their position arrangement is also important to ensure efficient boring.

Figure 4.52 Terra Invader rig used for deep drilling. *Courtesy of Herrenknecht AG.*

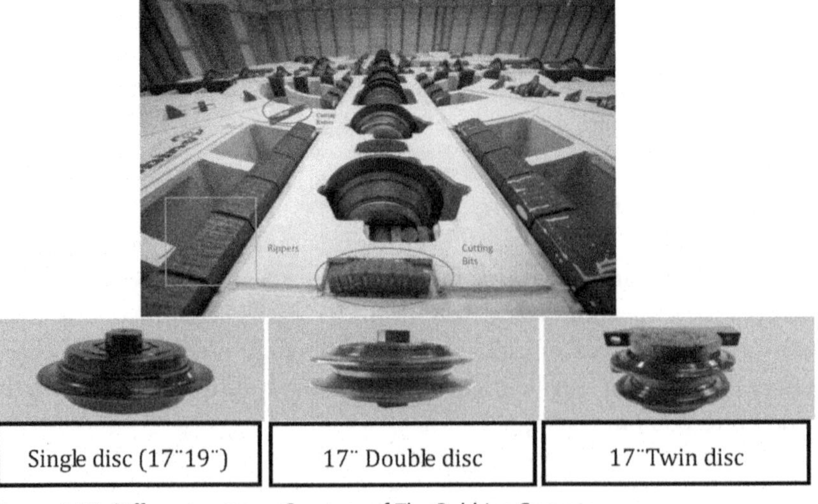

Figure 4.53 Different cutters. *Courtesy of The Robbins Company.*

Figure 4.54 Conveyor belt.

4.3.3 Muck Management

4.3.3.1 Muck Transportation

A muck removal system transfers the excavated material from the tunnel face to the surface (Fig. 4.54). The consistency of the excavated material can vary considerably depending on the type of shield used and the ground being excavated. This therefore requires different modes of muck transport (Maidl, Herrenknecht, Maidl, & Wehrmeyer, 2012) as follows:

- Dry removal for open shields, compressed air shields, and EPB shields operating in open or compressed air mode. A conveyor belt or chain conveyor together with locomotives are often used.
- Slurry removal for slurry shields. The support medium is also a transport medium, usually consisting of transportation pipes and hydraulic pumps.
- High-density slurry removal for EPB shields. A conveyor belt is usually used as the muck emerges in cylindrical chunks.
- The vertical muck transportation consists of pipes, conveyor belts, and buckets based on TBM type and tunnel depth.

There is no definite transportation mode for each shield type. The choice must therefore be made according to the project specifics (Fig. 4.55).

4.3.3.2 Muck Disposal

The excavated material should be recycled as much as possible on-site or nearby. Contaminated material that can be reused should be

Figure 4.55 Vertical hoisting box.

decontaminated, or if not, transported to treatment facilities or licensed landfill sites. Landfill site slope stability should also always be verified in advance.

4.3.4 Tunnel Lining

The lining secures the tunnel against the surrounding ground. It protects the tunnel from loads from natural movement of the ground and surface traffic and structures such as buildings. It can also waterproof the tunnel. There are three types of tunnel linings: pipe lining (pipe jacking), segmental lining, and *in situ* lining. This chapter focuses on segmental lining, the form that is most common in tunnels today. This type of lining is usually the support of choice in TBM tunnel construction.

4.3.4.1 Fabrication

Cast in mold (Fig. 4.56), segmental linings are made of reinforced concrete. The prevalent segments are 1.2–2 m wide and typically 25–50 cm thick. Segment production consists of the following steps:

1. Erect the steel cage and place it into the steel mold.
2. Pour concrete inside the mold.

Figure 4.56 Steel mold for segment.

3. Vibrate and cure the concrete.
4. Remove the segment from the mold.

As technologies have evolved, new reinforced materials such as synthetic fibers have partially replaced traditional steel bars. Fiber-reinforced concrete is more durable, economical, and environmentally friendly.

4.3.4.2 Storage

Segments are generally stored in a stacked arrangement (Fig. 4.57). Once the design strength is reached, segments are transported to the position where they must be in for erection inside the tunnel.

4.3.4.3 Transportation

A locomotive (Fig. 4.58) is used for segment transportation. While transporting, there are usually no more than three segments stacked.

4.3.4.4 Segment Erection

Several curved segments are assembled inside the tail of the TBM to form a complete ring. For medium-sized tunnels, normally five segments and a key constitute an entire ring. Installation usually starts with the invert segments, followed by the left- and right-side segments and then completed with the key segments. This is illustrated in

Figure 4.57 Stacked precast segments.

Figure 4.58 Transportation of segments within the tunnel.

Fig. 4.59. Segments are named according to their positions (Fig. 4.60) as key segments, adjacent segments, and standard segments. There are then two means of joining the segments:

1. Cross-joints. These segments are easy to assemble, though unevenness on the surface and accumulated errors are likely to occur.

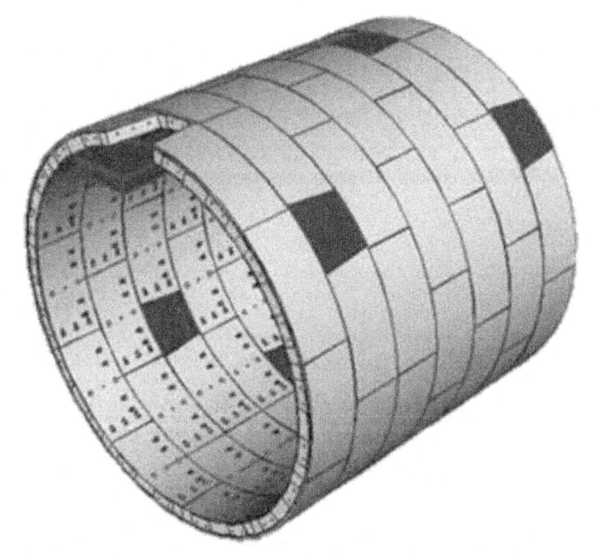

Figure 4.59 Segment installation.

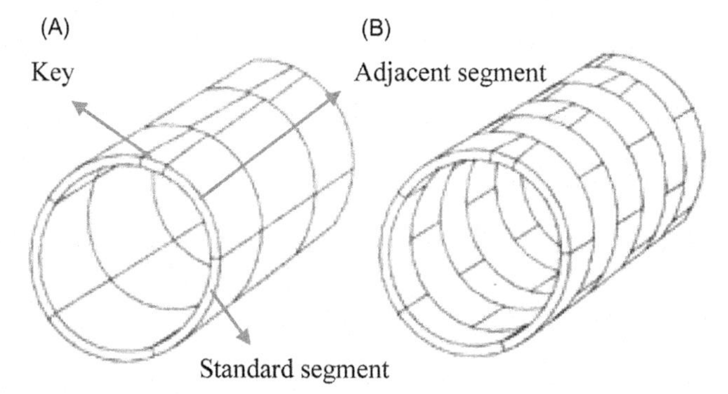

Figure 4.60 (A) Straight joint and (B) staggered joint.

2. T-joints. These segments are stacked. They also have better water tightness and rigidness, making them the preferred method of segmental lining.

4.3.5 Tailskin Seal

The tailskin seal protects the rear end of the shield against underground water and the surrounding ground. It is designed to provide a seal at the annular gap between the tunnel lining and the tailskin (Fig. 4.61).

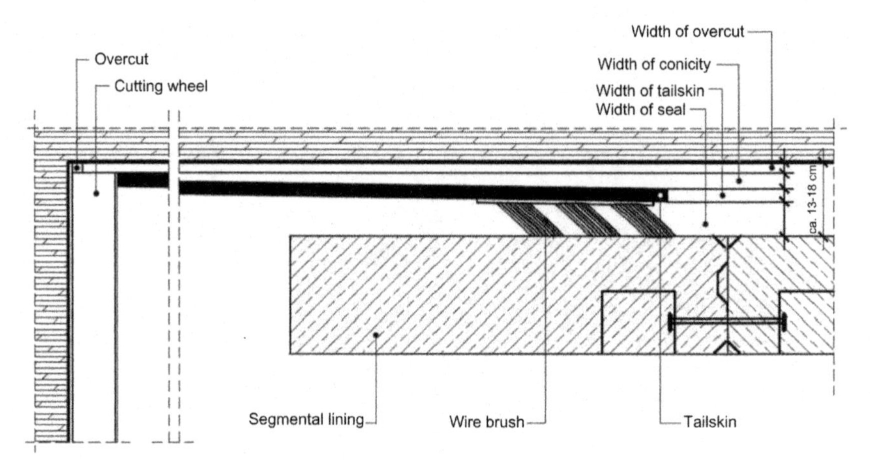

Figure 4.61 Detailed scheme of a TBM tailskin.

The seal is usually ensured by injecting tail seal greases between rows of wire brushes. The main objective of the grease is to prevent inflow from grouting, and soil or water inside the tunnel shield; in other words, it ensures the sealing of TBM while tunneling. Thus, the gap between the TBM shield and the segments will be filled by grease. Therefore, the tail-sealing grease system pressure should be higher than the grouting system when the TBM is excavating Nator Lubricants. Tail-sealing greases are high viscosity fluids. The grease performance is evaluated on the following main aspects according to the European Federation of Specialist Construction Chemicals and Concrete Systems (EFSCCS):

- Good resistance against water and grout pressure.
- Good antiwashout properties.
- Good wear protection for the brushes.
- Good pumping properties over a wide range of temperatures, which means that the apparent viscosity should be low.
- Good adhesion to concrete and metal.
- Good stability in storage and under pressure, which means that leaching and decomposition should be prevented no matter what the external environment is.
- No harmful effect to the sealing gaskets.

4.3.6 Grouting

The aim of grouting is to backfill the annular gap between the ground and the segment lining. To avoid settlement, it is important for the grouting process to commence as early as possible. Mortar and two-component grout (cement slurry and sodium silicate) are two common simultaneous grouting materials. There are two ways to grout: either through the tailskin or through prefabricated holes in the segments. The early way is simultaneous grouting, while the later way is called secondary grouting.

After simultaneous grouting, cavities can still exist. This can be due to primary grout settlement, for example. Secondary grouting is then needed and is usually done at a higher pressure than the ground pressure in order to pressurize and stabilize the ground. In the case of hard rock TBMs, the annular gap is backfilled with pea gravel instead of grout. This means that secondary grouting is required.

1. *Through the tailskin*

   Grouting directly through the tailskin (Fig. 4.62) is preferred in soft ground as there is a need to immediately fill the annular gap due to short standup times.

2. *Through grout holes in tunnel segments*

   Grouting can be done through existing holes in the lining segments (Fig. 4.63). These holes usually have a mechanism in the annular gap to retain the grouting material.

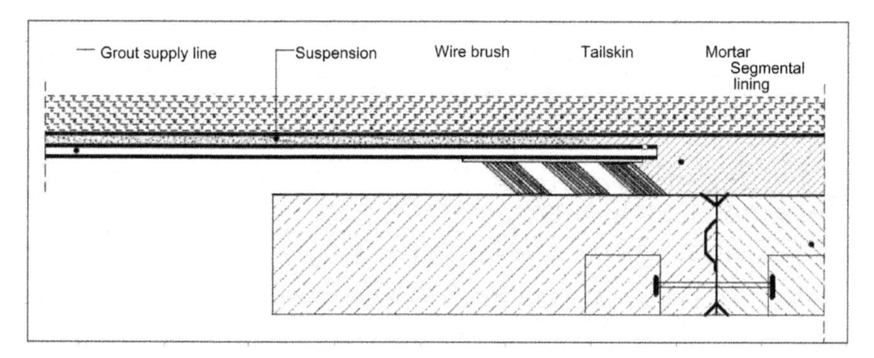

Figure 4.62 Grouting through the tailskin.

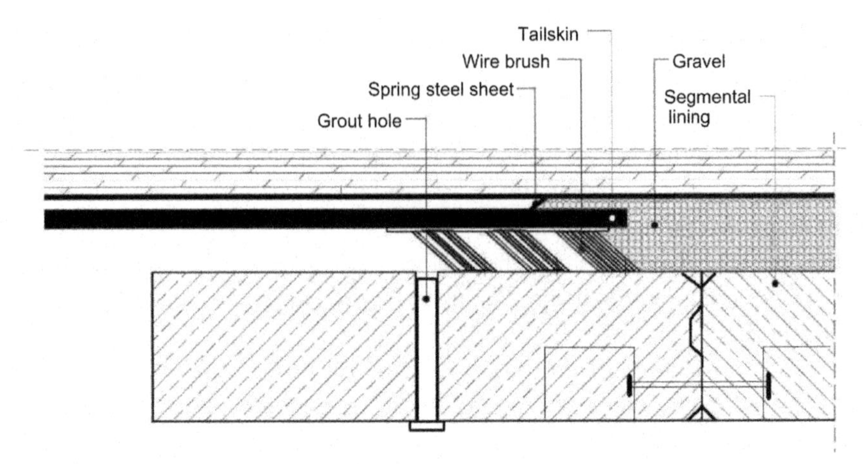

Figure 4.63 Grouting through grout holes in the lining segments.

4.3.7 Surveying and Guidance Systems

For a TBM tunnel, it is of paramount importance to have reliable surveying and real-time guidance systems to guarantee the excavation follows the predefined alignment. Depending on the required accuracy of the machine drive, machine type, and tunnel alignment, measuring data is obtained using either manual surveying or automatic sensors.

To help the operator steer the TBM, tunnel guidance systems track its position and direction (Lee, 2007) as follows:

- A laser station determines the three-dimensional (3D) position continuously.
- The video target determines the twist in the XY plane.
- Inclinometer 1 determines the roll in the XZ plane.
- Inclinometer 2 determines the tilt in the YZ plane.

4.3.8 Comparison of Tunnel Boring Machine and Conventional Excavation

The choice between drill-and-blast and TBM tunneling is still debated to this day. This is because the choice is not always evident. Table 4.2 shows some of the differences between the two methods.

The emergence of mechanized tunneling has made it possible to reach considerably higher advance rates compared to traditional

Table 4.2 Main comparison of TBM and drill-and-blast excavation methods

Feature	Drill-and-blast	TBM
Shape	Any shape	Mostly round
Applicable ground	Rock	Hard rock/soft soil
Overbreak	Inevitable	Eliminated
Support	Temporary in situ support needed for weak rocks	Lining
Operation	Dangerous, unpleasant working environment	Safer and can be cheaper in terms of lifetime costs
Crews	Skills required	Easily trained
Access structure	Shafts and adits to open multiple headings	Launch and arrival shaft
Influence	Noise, vibration	Environmentally friendly
Grouts	Shotcrete if necessary	Synchronous grouting
Flexibility	Site change during the same drive	Hard to change on-site

tunneling. But it is not suitable for all ground conditions and projects. The TBM method requires more comprehensive and thorough geological investigation, mapping, and testing during the planning stage. This is to evaluate the feasibility of using a TBM, and if so, which type should be used.

The costs of the two methods is mainly related to the length of the tunnel. For longer tunnels, TBMs are more economical as the high investment of the machine cost is distributed over the longer distance. The cross-point of the cost of the two methods is about in the range of 10–15 km according to past experience.

To overcome some of the disadvantages of the drill-and-blast method, the Kajima Corporation of Japan has developed a new type of tunnel machine as shown in Fig. 4.64. This machine offers advantages of both tunneling methods.

4.4 IMMERSED TUNNELS

Immersed tunnels are made of prefabricated sections and are widely used to cross shallow water (Fig. 4.65). Tunnel elements are fabricated elsewhere and then towed to the tunnel location before being joined together and buried permanently.

Figure 4.64 NATBM Kajima Corporation of Japan.

Figure 4.65 Image of an immersed tunnel.

4.4.1 Types of Immersed Tunnels

Immersed tunnels can be divided into two types depending on the method of fabrication used (ITA Working Group n°11 (ITA WG11), 2016b).

- Steel-concrete composite tunnels: stiffened steel plates are used that act together with an inner concrete filling. Tunnel elements are usually fabricated in yards and then the concrete is poured.
- Reinforced concrete tunnels: concrete is reinforced by steel bars or prestressing cables. Tunnel elements are cast in a dry basin before being flooded to be transported.

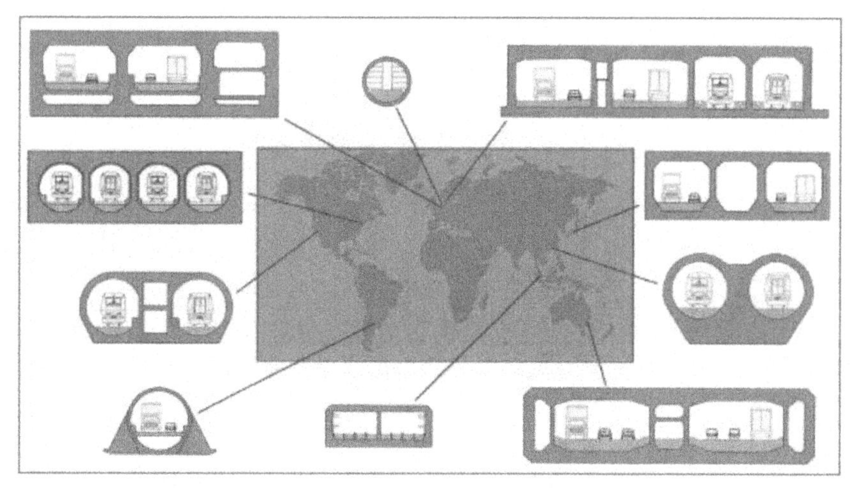

Figure 4.66 Worldwide examples of cross sections of immersed tunnels (ITA WG11, 2016a).

In Europe, Southeast Asia, and Australia most immersed tunnels use reinforced concrete. But in Japan, both types of tunnels are used in approximately equal numbers, and in North America mostly steel tunnels exist (Fig. 4.66).

The different kinds of immersed tunnels will be discussed in this section. Steel tunnels are single-immersed shell, double-immersed shell, or sandwich. There are also concrete-immersed tunnels.

4.4.1.1 Single-Shell Steel Tunnel

For single-shell steel immersed tunnels, the external structural shell plate works compositely with the inner reinforced concrete. As there is no external concrete, the steel plates are protected against corrosion. The Hong Kong cross-harbor tunnel (Fig. 4.67) is an example of such a structure.

For single-shell tunnels, leaks in the steel shell may be difficult to find and seal. Although subdividing the surface into smaller panels by using ribs will improve the chances of sealing a leak, great care and considerable testing is required to ensure that the welds are defect free. The risks of permanent leakage can therefore be higher in single-shell immersed tunnels than in other types.

Figure 4.67 Hong Kong cross-harbor tunnel elements nearly ready for side launching (FHWA, 2009).

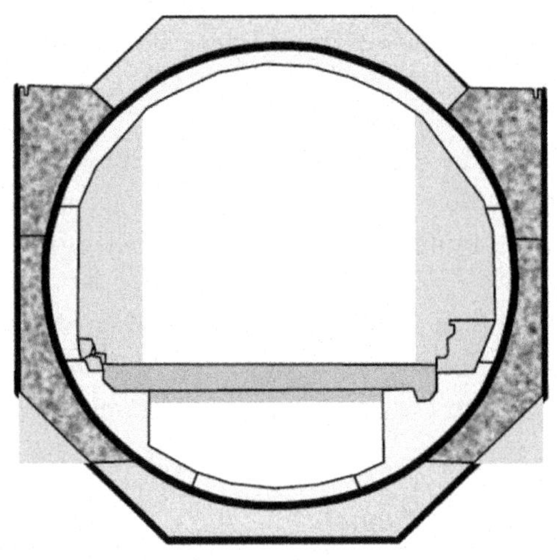

Figure 4.68 Double-shell cross section for the second Hampton road tunnel, Virginia (FHWA, 2009).

4.4.1.2 Double-Shell Steel Tunnel

For double-shell tunnels, an internal structural shell acts together with concrete placed within it. Another structural element is the top and invert concrete which are placed outside the structural shell plate.

A second steel shell is constructed to act as a frame for ballast concrete at the sides. The interior structural shell is therefore protected by external concrete that is placed with nonstructural steel plates (FHWA, 2009). Fig. 4.68 shows the cross section of the second Hampton roads tunnel in Virginia. The thick black lines correspond to steel plates.

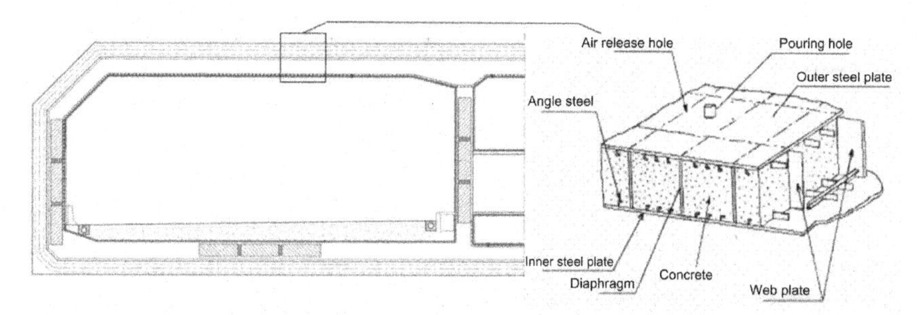

Figure 4.69 Scheme of a sandwich construction. *Modified from US Department of Transportation Federal Highway Administration. (2009).* Technical manual for design and construction of road tunnels—Civil elements. *Publication no. FHWA-NHI-10-034.*

4.4.1.3 Sandwich Construction

Sandwich construction consists of an inner and outer steel shell that act together with an unreinforced structural concrete layer that is sandwiched between them (Fig. 4.69). For a successful composite action, the inner surfaces of the steel shells must have connectors. The result is that steel shells carry tension loads while the concrete bears compression loads (ITA WG11, 2016a).

4.4.1.4 Concrete-Immersed Tunnels

The advantage of concrete is that it is durable and can be molded to any shape. Being heavy, concrete tunnel elements are usually constructed in dry basins near the project-site fabrication factory (Figs. 4.70 and 4.71) before being flooded and towed into position.

In concrete structures, uneven shrinkage is the major cause of cracking (Bickel, Kuesel, & King, 1996). It is therefore important to implement a control crack method when using concrete-immersed tunnels.

4.4.2 Construction Method

The construction of an immersed tunnel consists of several steps. These are namely trench excavation, foundation preparation, tunnel element fabrication, transportation and handling of tunnel elements, element lowering, element placing, and backfilling.

1. *Trench excavation*

   Trench excavation means excavating an open trench in the bed of the body of water being crossed. This excavation is usually carried out with a clamshell dredger (Fig. 4.72).

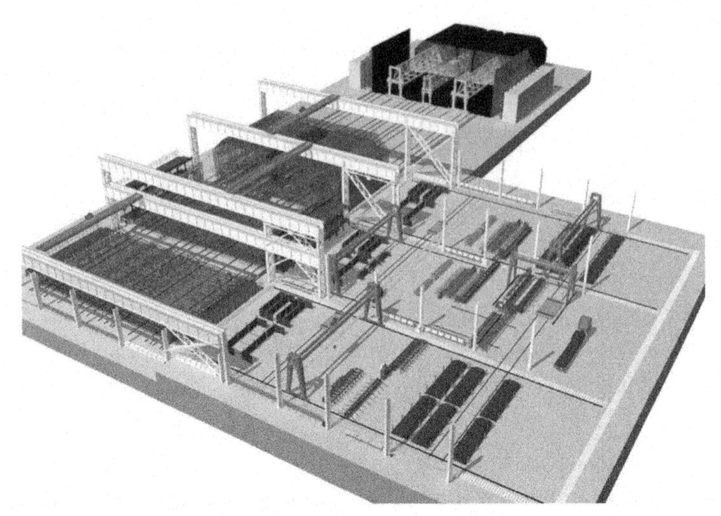

Figure 4.70 Hong Kong-Zhuhai-Macau Bridge (HMZB) and Tunnel System. One of the world's longest undersea vehicular tunnel projects.

Figure 4.71 Full section of the hydraulic automated formwork system HMZB tunnel.

2. *Foundation preparation*

   The foundation should be installed once the trench excavation is completed. Most countries worldwide are applying a combination of foundation techniques for immersed tube elements. This is because geological, environmental, and economic conditions vary.

   The popular foundation types include sand injection, grouts injection, pipe foundation and ground foundation, etc. (Fig. 4.73).

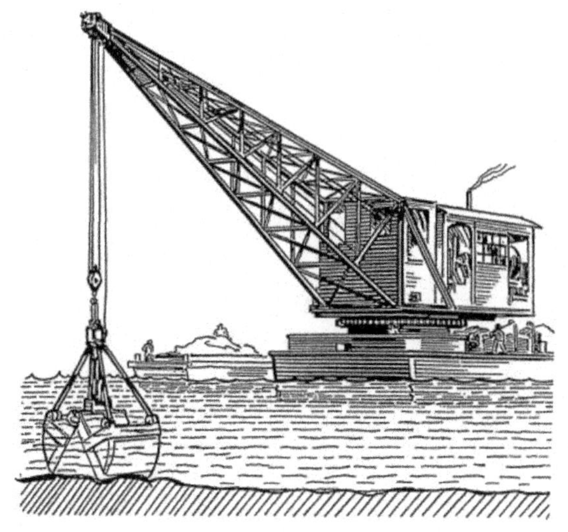

Figure 4.72 Clamshell dredger (Pearson Scott Foresman, 2009).

Table 4.3 gives a comparison of three main types of foundation used in four projects in China. From this table we can see that the pile formation has little settlement but with high cost, whereas sand flow is cheapest with moderate settlement results.

3. *Tunnel element fabrication*

   Tunnel elements are usually fabricated off-site at a fabrication factory, shipyard, or in dry docks. Elements constructed on launching ways are slid into the water (in the same way as ship launching). To make the elements watertight, they are closed by bulkheads. The bulkheads are set at a nominal distance from the end of the element, resulting in a small space at the ends of the adjoining sections. This space is filled with water and dewatered after the connection with the previous element is made (Fig. 4.74).

4. *Transportation and handling*

   After fabrication and launching, the elements are towed into position over the excavated trench (Fig. 4.75). To avoid damaging them, the elements are designed to withstand the thermal and hydrostatic effects that they are later subjected to. Transportation should also not take place during adverse weather conditions to avoid exceeding operational limits.

   It is expected that the wind speed will be ≤ 10 m/s, wave height ≤ 0.5 m, water velocity ≤ 1.0 m/s, and visibility ≥ 1000 m during transportation.

1. Sand injection *2. Pipe foundation*

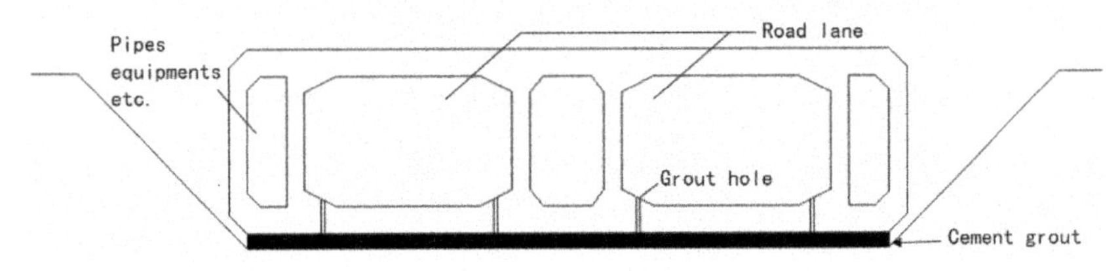

3. Grouts injection

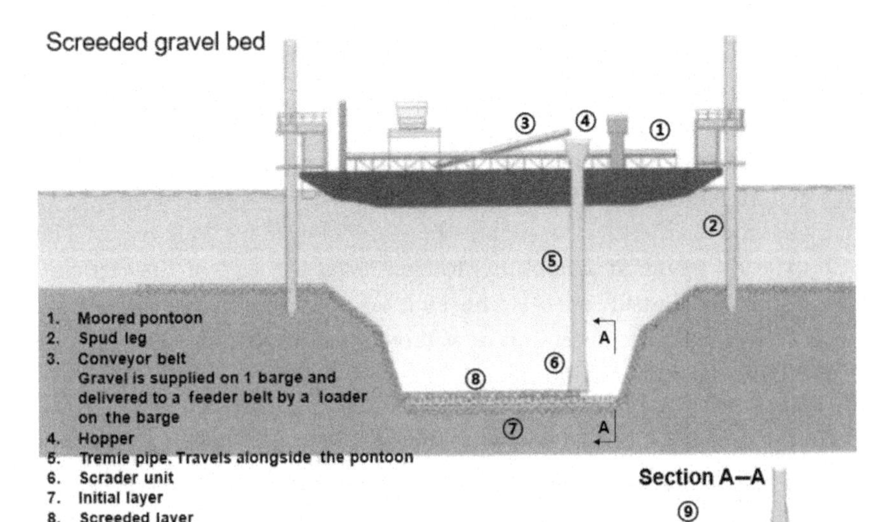

4. Gravel foundation

Figure 4.73 Foundation types.

Table 4.3 Comparison of four Chinese immersed tube tunnel foundations

Project	Yong Jiang project	Zhu Jiang project	Chang Hong project	Outer ring project
Start year	1987	1988	1999	1999
Purchasing method	Traditional	Traditional	Design-build	Design-build
Construction duration (years)	8	3	2	4
Hydrogeology	Serious sedimentation and soft ground	Highly weathered rock and clay	Serious sedimentation and soft ground	Serious sedimentation and soft ground
Foundation type	Grouting	Sand flow	Pile	Sand flow
Accumulative settlement (cm)	27.5	4–5	2	15
Foundation cost million (USD)	0.36	5.00	20.00	13.50
Contract sum million (USD)	25	1.73	3.72	10.99
Ratio of foundation/contract (%)	1.4	2.89	5	1.22
Tube length (m)	420	457	400	736
Tube width (m)	11.9	33	22.8	43

Figure 4.74 Installation of a continuous rubber gasket. *Courtesy of Hong Kong Zhuhai Macao Bridge Authority.*

Figure 4.75 Tunnel elements in transportation.

5. *Lowering and placing*

Placing an element can be a challenge as the water current must be considered. After arriving at its destination, the tunnel element is lowered onto preprepared foundations that lie in a trench on the sea bed. This is typically done by purpose-built catamarans, pontoons, or even cranes (FHWA, 2009). If the element does not have enough negative buoyancy, a temporary ballast system can be used to help lower it to make the top side of the board only 10–25 cm higher than the water level.

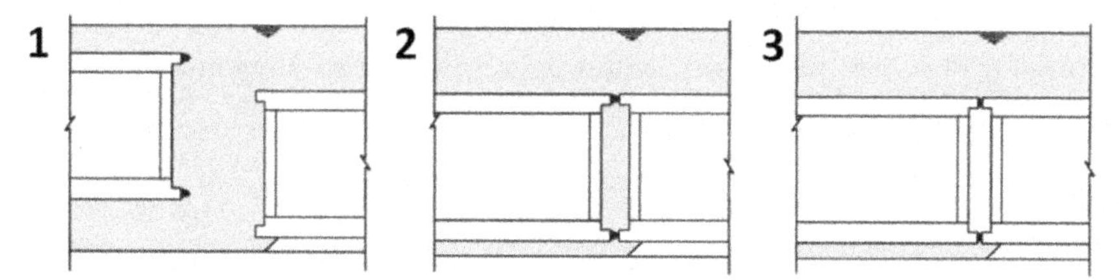

Figure 4.76 Scheme of the connection of two elements (1. Lowering of the element; 2. Placement of the element; 3. Connection of the elements and pumping out water) (Lunniss & Baber, 2013).

Instrumentation is installed inside the tunnel elements during fabrication to enable long-term settlement monitoring. After backfilling, monthly recordings should be carried out until settlement becomes negligible.

After placing the element in its position, a connection is made between it and the end face of the previously placed element (Fig. 4.76, scheme 1). The water trapped (Fig. 4.75, scheme 2) in the joint between the two elements is pumped out (Fig. 4.76, scheme 3).

Locking fill is placed in the trench, to a minimum of half the height of an element (FHWA, 2009). This secures tunnel elements in position after all tunnel elements have been connected. The filling is usually sand or a coarser material.

Ordinary backfilling of the trench to a depth of at least 1.5 m above the tube is done with firm material. In long tunnels, soil excavated from the trench can also be used as backfill over tubes that have been placed in other parts. Backfill should be free of clay balls and be chemically inert (FHWA, 2009).

4.4.3 Waterproofing

Immersed tunnels have few in situ joints, which is an advantage over bored tunnels when it comes to waterproofing. The water tightness of immersed tunnels depends on (ITA WG11, 2016a) several factors:

- for steel shell tunnels: on weld quality, in situ joint quality, and flexible joint quality and
- for concrete tunnels: on joint quality, the absence of full-depth cracks in the concrete, and waterproofing quality.

Some concrete tunnels come with enveloping watertight membranes. This also provides protection against aggressive chemical agents for structural concrete.

4.4.4 Immersion Joints

Concerning immersion joints, before transportation tunnel elements are fitted with a continuous rubber gasket (Fig. 4.77) around their external perimeter.

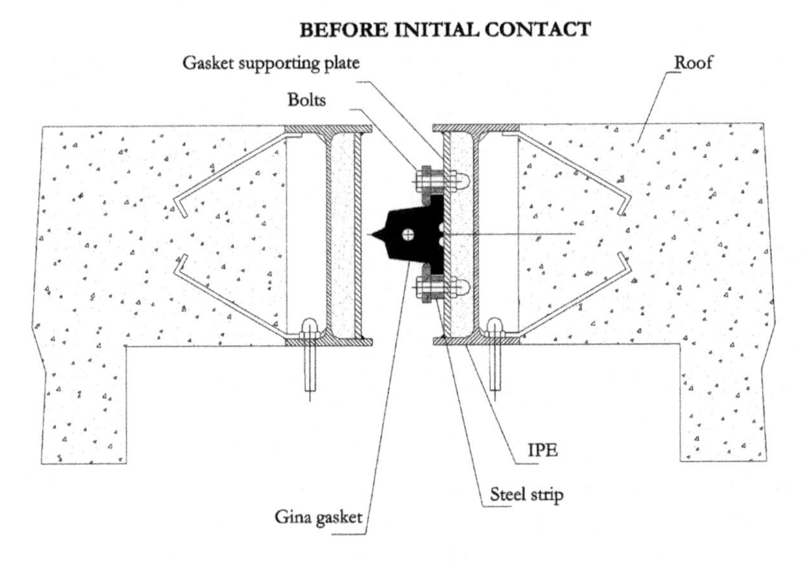

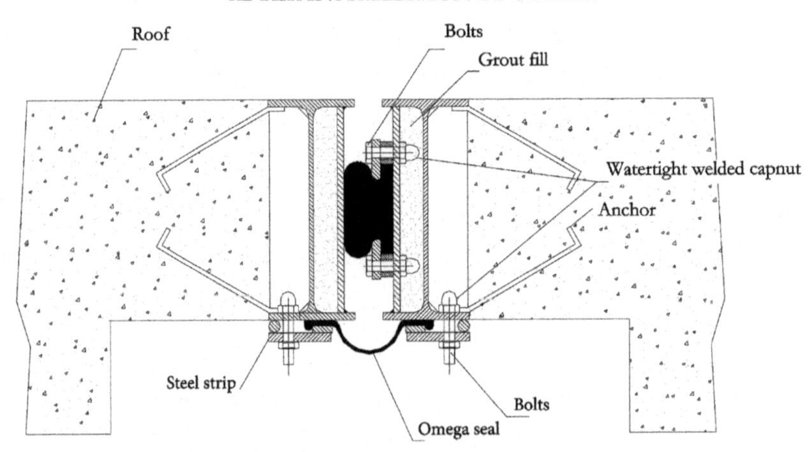

Figure 4.77 Cross section of an immersion joint (Trelleborg, n.d.).

The Gina and Omega combination is often adopted for concrete-immersed tunnels as it has proved to be a robust and safe solution (Lunniss and Baber, 2013). In the immersion stage when the elements are assembled, the Gina gasket's soft nose (Fig. 4.77, upper scheme) is pushed against the end face of the opposite tunnel element. This provides the initial seal that enables the dewatering of the space between the elements (Fig. 4.76, scheme 3). Pressure difference will cause the tunnel element to move closer to the previously immersed one, causing further compression of the Gina gasket (Lunniss and Baber, 2013).

As secondary protection, an Omega seal is installed (Fig. 4.77, lower scheme). This element allows axial and radial freedom of the two-bridged structures (Trelleborg, 2009) while preventing relative displacement between them.

4.4.5 Advantages and Drawbacks of Immersed Tunnels

If water is not deeper than 30 m, immersed tunnels usually offer a feasible alternative to bored tunnels at a comparable price, in addition to other advantages:

1. Almost any cross section can be constructed. This makes immersed tunnels attractive for different means of transportation (Fig. 4.78).
2. Immersed tunnels can be installed in most types of ground. This makes them more flexible than bored tunnels for which soft alluvial materials can be a problem.
3. The three stages of immersed tunnel construction (dredging, tunnel element construction, and tunnel installation) can be done simultaneously.

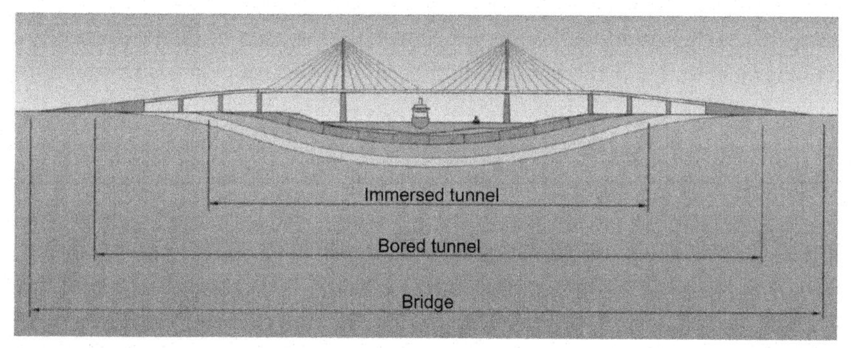

Figure 4.78 Sketch comparing the length of different civil engineering structures to cross a waterway (ITA WG11, 2016a).

Immersed tunnels can therefore be completed in a shorter time than bored tunnels.

4. Immersed tunnels can be placed in shallow areas (Fig. 4.78)
5. Immersed tunnel section sites can be longer than those in a TBM tunnel. Thus, an eight-lane road tunnel is possible in an immersed tunnel.
6. The section site can be varied in an immersed tunnel.

However, immersed tunnels also have disadvantages:

1. Immersed tunnels may influence water transportation during section transportation and lowering stages.
2. The selection of suitable sites for constructing the tunnel elements can be difficult.
3. The sea bed must not be too soft or unstable so the trench can be kept open.
4. The site must be reasonably free from rapid deposition of fluid silts, which can alter the density of water in the trench and affect the buoyancy balance of the tube.

4.5 BUND TUNNEL CASE STUDY: KEY PROTECTION TECHNIQUES ADOPTED AND ANALYSIS OF INFLUENCE ON ADJACENT BUILDINGS DUE TO CONSTRUCTION

Source: Bai, Y., Yang, Z., & Jiang, Z. (2014). Key protection techniques adopted and analysis of influence on adjacent buildings due to the Bund Tunnel construction. *Tunnelling and Underground Space Technology, 41*, 24–34.

4.5.1 Introduction

The Bund Tunnel is 14.27 m in diameter. It is the first application of a superdiameter earth pressure balanced shield (EPBS) in China. There are many historical buildings along the construction line, and the minimum horizontal distance from the building to the tunnel side varies from 1.7 to 30 m. Considering the importance of these historical buildings and the complicated construction processes, it is essential to adopt effective protection techniques to ensure safety during the tunnel construction. Three kinds of protection techniques are presented in this paper. Firstly, an underground cutoff wall built by bored piles is used to separate the buildings and tunnel when the minimum horizontal distance from the building to the tunnel side is less than 5 m. Secondly, the grouting reinforcement technique is adopted when the minimum clear distance is between 5 and 10 m. Finally, if the minimum clear distance is larger than 10 m, the optimized construction parameters are selected to reduce the influence induced by the EPBS excavation. The deformations of some typical buildings are monitored. The results of this project will be a useful reference for similar future projects.

To relieve the traffic pressure of the Shanghai business district and improve the surface environment of the Bund area, a comprehensive traffic renovation project was proposed in 1992, and the Bund Tunnel was an important part of the project. The project started its construction in 2007, and was open to traffic on March 28, 2010. The largest EPBS of 14.27 m diameter was used. It was the first application of a superlarge-diameter EPBS in China.

Fig. 4.79 illustrates the general layout of the Bund Tunnel. The tunnel begins at South Zhongshan Road in the south and ends at Haining Road in the north. Its total length is 3.3 km. From Haining Road to the north of East Yan'an Road, the tunnel was built with EPBS, whereas in the southern part of the East Yan'an Road, the tunnel was built using the cut-and-cover method.

The detailed layout of the section excavated by EPBS is plotted in Fig. 4.80. It is obvious that the construction environment of the tunnel is very complicated, as the tunnel has to pass several existing traffic tunnels and underground structures either above or below its position. For instance, it passes above the Line 2 metro, and passes beneath the East Nanjing Road underpass, the East Beijing Road underpass, and the pile foundation of the century old Waibai Du Bridge. There are also several dozen historical buildings with different architectural styles including Baroque, Gothic, Rococo, and traditional Chinese architecture adjacent to the Bund Tunnel. These historical buildings are not only part of the city's the cultural heritage but also the icons of the city.

The soil layer in Shanghai is soft soil. Table 4.4 presents the mechanical properties of the ground layer around the Bund Tunnel.

The adjacent buildings and structures were vulnerable to deformation and possible damage due to tunnel excavation. Therefore, it was essential to predict the settlement during the tunnel construction. The methods of predicting the settlements can be categorized into four types: empirical methods (Attewell & Woodman, 1982; Loganathan & Poulos, 1998; O'Reilly & New, 1982; Peck, 1969; Toraño et al., 2006), analytical methods (Chou & Bobet, 2002; González & Sagaseta, 2001; Hisatake, 2011; Rowe & Kack, 1983; Sagaseta, 1987; Verruijt & Booker, 1996), model test methods (Farrell, 2010), and numerical methods (Lee & Rowe, 1990a, 1990b).

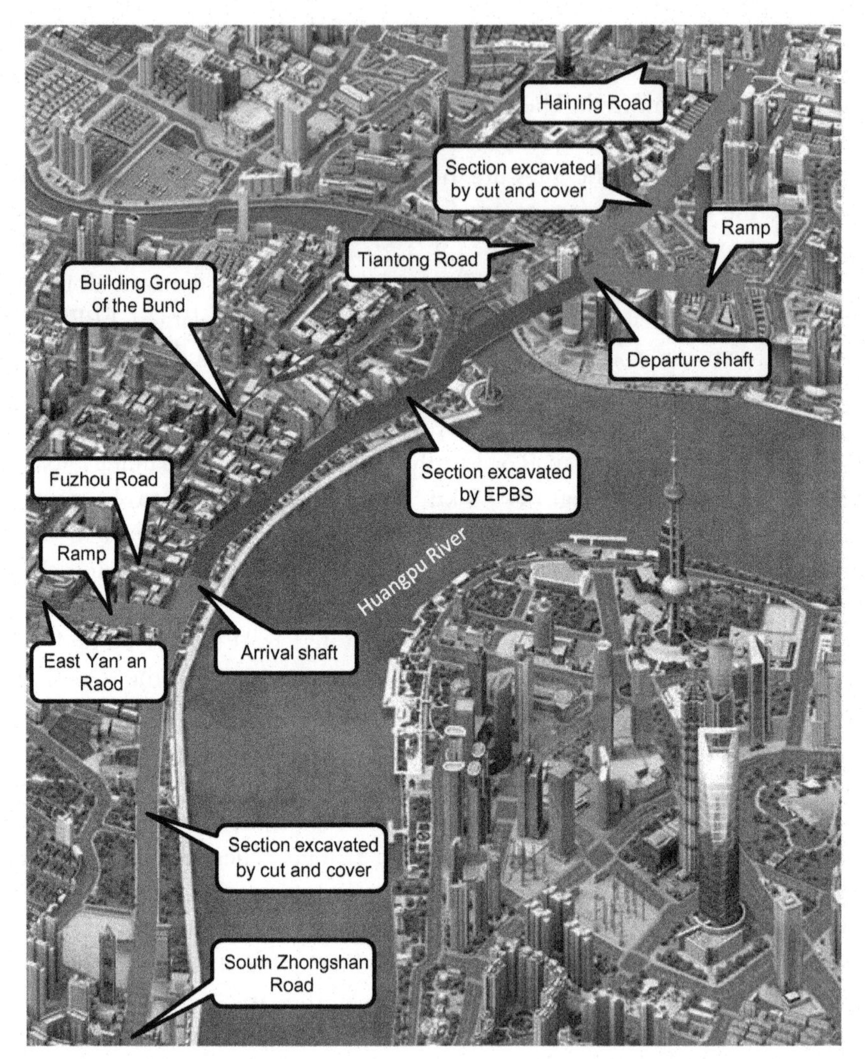

Figure 4.79 General layout of the Bund Tunnel.

In order to ensure the safety of adjacent buildings and structures during the tunneling, it was necessary to take some protective measures, including setting up a monitoring system for the adjacent buildings, and analyzing the influence of the tunnel construction. In this chapter, three types of protective measures and their applications are illustrated. The deformation of the Pujiang Hotel, the Shanghai Mansion, and other historical buildings at the Bund were monitored. Furthermore, the principle and effect of each protection measure is described.

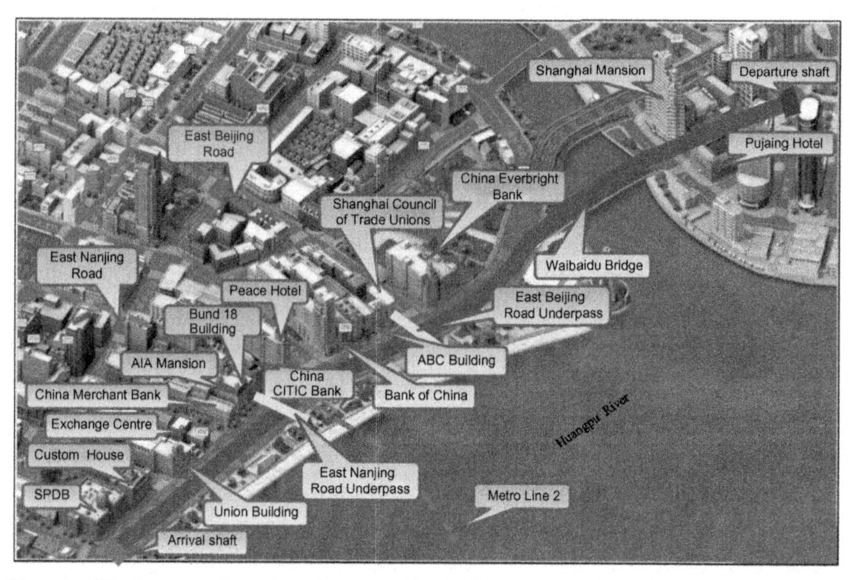

Figure 4.80 Layout of section excavated by EPBS.

Table 4.4 Mechanical properties of the ground

List of layer	Name of ground	Compression modulus, E (MPa)	Cohesion, C (kPa)	Inner friction angle, ϕ (degrees)
①	Fill	10	5	10
$②_0$	Silt soil	9.55	4	31
$②_1$	Silty clay	6.04	18	11
$②_{3-1}$	Muckysilty clay	8.04	10	25
$②_{3-2}$	Clayey silt with silty clay	9.03	7	28.5
③	Muckysilty clay	3.44	15	13.5
④	Silty clay	2.46	14	12
$⑤_1$	Gray clay	4	10	22.5
$⑤_3$	Silty clay with clayey silt	16.7	17	21
$⑤_4$	Silty clay	15.4	7	27.5

4.5.2 Measures Taken to Protect Adjacent Buildings and Structures

There are a number of buildings and structures near the tunnel. Most of the buildings were built in the early 20th century, and almost all of them were included in the National Historical Protection Building Group. The buildings had various degrees of deterioration due to decades of weathering and ground settlement. Therefore, it was essential to take measures to protect the buildings. The buildings were divided into three categories based on the minimum horizontal distance to the tunnel side (the dimension d in Fig. 4.81), and different kinds of protective measures were selected after a comprehensive consideration of safety, economics, and schedule.

1. *Full-section isolation measures with bored piles*

   During the tunnel construction, the shield was expected to pass very close to some buildings. Normal measures including controlling the soil chamber pressure, simultaneous grouting, and reducing driving speed were insufficient to meet the protection aims. The most effective way to substantially reduce the influence was to isolate the buildings from the effects of tunneling.

   In situ mixing piles, static-driven steel piles, sheet piles, and jet grouting piles are common isolation methods used in China. Each type of methodology has its disadvantages when used for protecting buildings within 5 m ($d<5$ m). The strength of the in situ mixing piles is too low to isolate the tunneling disturbance from buildings. Sheet piles have an intense extrusion effect on the surrounding soil due to its vibratory nature, which may lead to liquefaction of the sandy layer. Jet grouting piles often form in irregular columns with various strengths, which may not guarantee the isolation.

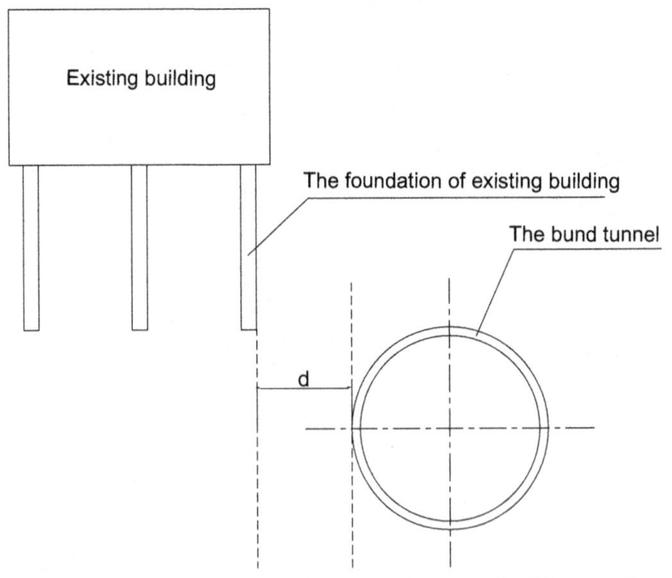

Figure 4.81 Cross section with relative location of existing buildings and tunnel.

Compared with the other described measures, bored piles have advantages such as causing only slight vibration, requiring small working area, and leaving little squeezing in ground. Therefore, this method can effectively protect the adjacent buildings. Much research has been carried out on the response of bored piles during the tunneling process. For example, Chen et al. (1999) analyzed the lateral and axial response of piles caused by tunneling with a two-stage approach. Xiang et al. (2008) introduced a protection scheme that included two small cross-sectional drifts with shotcrete linings and piles used to separate the negative effects of tunneling on two nearby high-rise residential buildings and analyzed the protection effect with monitoring data. Xu and Poulos (2001) carried out analysis of vertical piles subjected to passive loadings including the tunneling process with a 3D coupled boundary element approach. Pang et al. (2005) set up an FE model with a feedback process to analyze the behavior of one of the instrumented pile groups during shield tunneling and discussed the effects of tunnel advancement on adjacent pile foundations. Lee (2012) conducted 3D elastoplastic numerical analyses of the response of piles during tunneling in weak weathered rock.

The bored piles were constructed with a fully cased drill rig including outer casing and inner screws. This type of machine has been widely used in underground engineering but was first used to protect historical buildings in Shanghai.

2. *Grouting reinforcement measures*

For the condition of *d* (Fig. 4.82) between 5 and 10 m, it was neither necessary nor economical to use contiguous bored piles. The

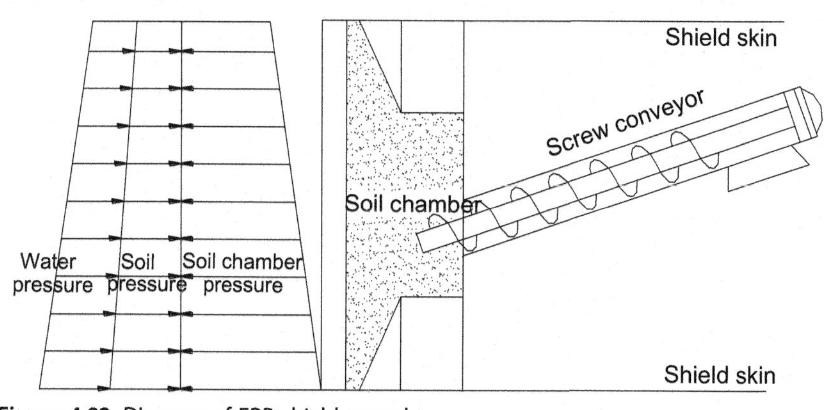

Figure 4.82 Diagram of EPB shield tunnel.

grouting technique was chosen because of advantages such as simple construction equipment at low cost, slight vibration, and minimal noise. The grouting technique has been widely used to reduce potentially damaging movements due to tunnel construction (Harris et al., 1994; Wisser et al., 2005).

3. *Regulating construction parameters*

The above two types of protection measures were aimed at building protection. But it was also important to reduce the shield driving effects. For buildings with over 10 m distance from the tunnel boundary, the most economical and effective way to protect the buildings was to carefully control the construction parameters. Xu et al. (2011) presented the relationships between the machine operation parameters and their disturbance to surrounding soil mass with laboratory model tests results. Fargnoli et al. (2013) investigated the influence of different excavation parameters on the subsidence trough. Chakeri et al. (2013) discussed the effects and important factors on surface settlement caused by metro tunnel excavated by EPBS. Key construction parameters include soil chamber pressure, driving speed, parameters of synchronized grouting, and shield posture.

- Selection of soil chamber pressure

  During EPB shield driving, the soil chamber pressure is used to maintain the excavation face's stability. Fig. 4.82 illustrates the relationship between the water-earth pressure and the chamber pressure. If the soil chamber pressure were less than the water-earth pressure, the surface in front of the shield would exhibit significant settlement. Conversely, the forward ground upheaval would appear. Both excessive settlement and uplift of the ground would cause disturbance to the adjacent buildings. During tunneling, the shield crosses multilayered soil, and the soil chamber pressure is regulated according to the value of the soil pressure sensors installed in the soil chamber.

  A 3D numerical model was set up to study the relationship between the settlement of the ground and the soil chamber pressure. Fig. 4.83 shows the scope of the model was determined to be 50 m in the vertical dimension, 120 m in the transverse dimension, and 80 m in the length. The soil layers of ②$_{3-1}$, ④, and ⑤$_3$ were simulated.

  Fig. 4.84 presents the relationship between the soil chamber pressure and the displacement of the forward ground. The K_0 represents

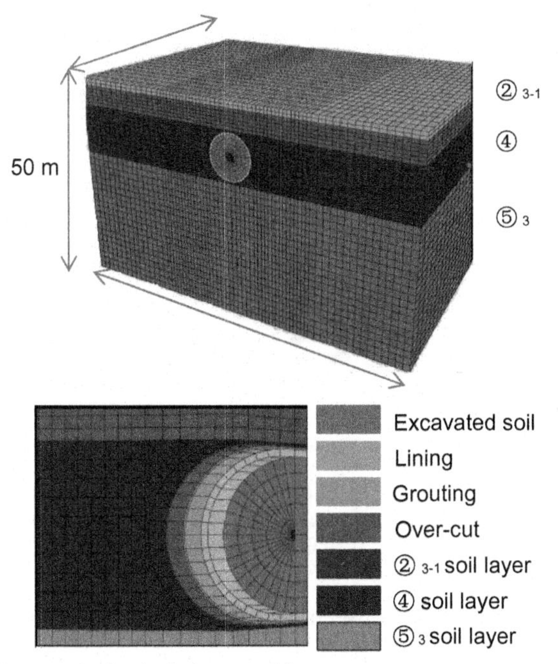

Figure 4.83 3D numerical calculation model.

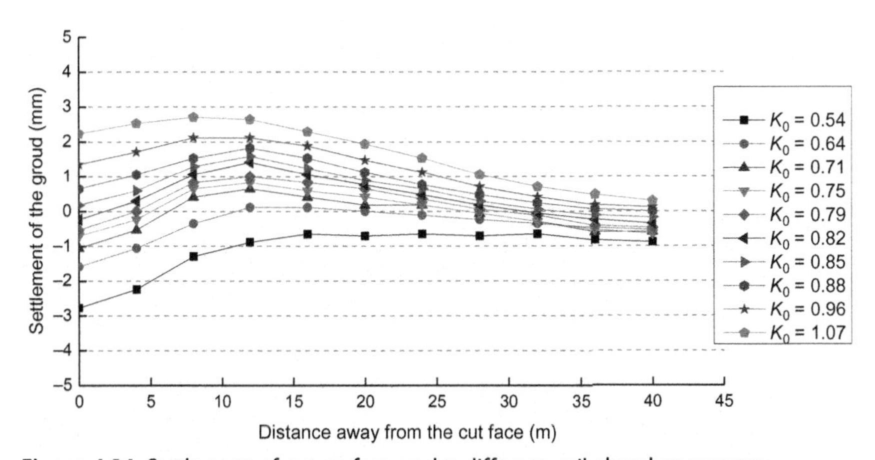

Figure 4.84 Settlement of cut surface under different soil chamber pressure.

the ratio of the soil chamber pressure to the vertical earth pressure at the center of the excavation face. The graphic also shows a linear change of settlement related to the soil chamber pressure. Thus, in order to keep the deformation within 1–2 mm, a reasonable value range of K_0 is 0.75–0.87 (Fig. 4.85).

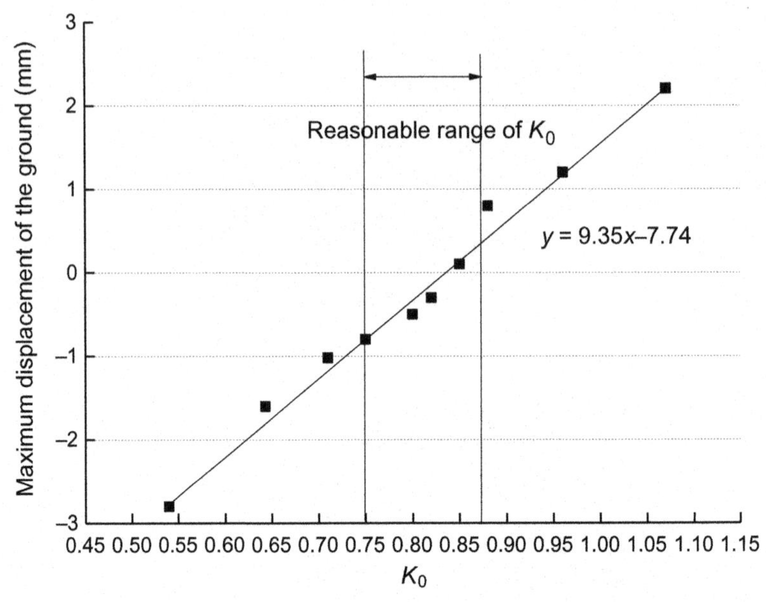

Figure 4.85 The relationship between the cut surface's displacement and the soil chamber pressure.

- Control of the tunneling speed

  After launching the shield driving, stopping should be avoided if at all possible. An extended pause will likely cause the pressure in the soil chamber to drop due to possible retraction of the jacks.
- Synchronized grouting

  After installing the segments, grouts are injected into the ground through six grouting holes distributed evenly on the periphery of the shield tail. The grouting pressure, the flow rate, and the grouting quality are the primary parameters of synchronized grouting. The grouting pressure varies with the depth of the tunnel and properties of the soil layer. The tunnel depth was within 10–16 m in the section adjacent to the historical building group, and the shield mainly passed through the ④ and $⑤_1$ soil layers. To minimize the effect on adjacent buildings, the grouting pressure was set between 0.41 and 0.64 MPa, and the grouting pressure of grouting holes on the lower part of the segmental lining was larger than that of grouting holes on the upper part. The flow rate was approximately 0.015 m^3/min. To ensure the quality of the grout, the slump and shear strength of the grout were tested every day.

- Control of the shield posture

During the shield driving process, the shield posture control is another effective way to reduce the ground disturbance. The horizontal deviation and elevation deviation should be controlled within ±70 mm, and the deviation should be rectified in each driving to avoid sharp rectification.

4.5.3 Analysis of Influence on the Adjacent Buildings Due to Tunnel Construction

4.5.3.1 Analysis of Influence on the Pujiang Hotel

The Pujiang Hotel was built in 1860 in the Baroque style of the Victorian period, and it was retrofitted into neoclassical architecture in 1907. The main structure is comprised of timber, brick, and concrete, and a strip foundation and short piles were used. The separation between the hotel and the tunnel is shown in Fig. 4.86. The horizontal distance from the foundation to the external boundary of the tunnel is in the range of 1.7–4.5 m.

1. *Deformation analysis of the Pujiang Hotel without the bored piles*

 The Pujiang Hotel had been in service for approximately 150 years and the structure has deteriorated over the years. Some existing damage to the Pujiang Hotel included foundation settlement, wall cracking, and inclination of the building. It was essential to analyze the deformation of the building in order to take effective protective measures.

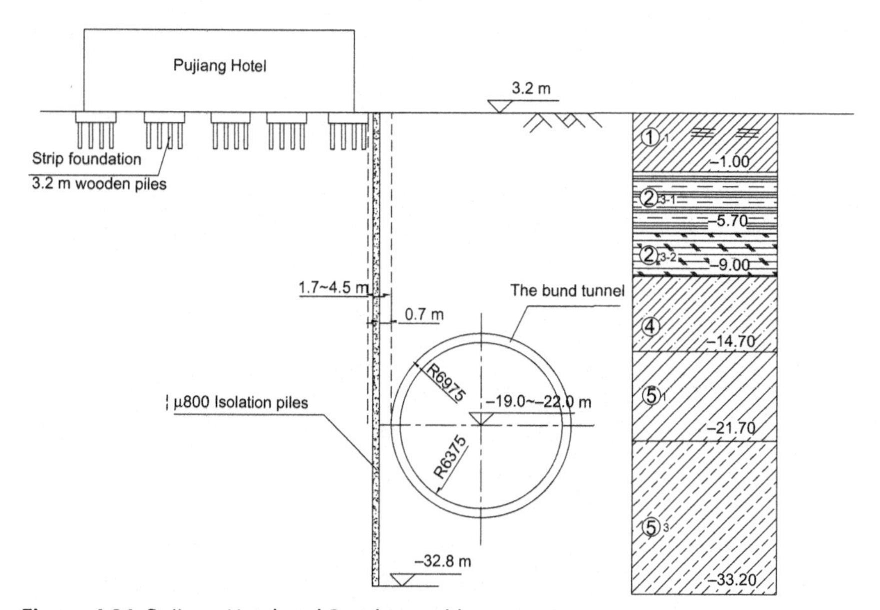

Figure 4.86 Pujiang Hotel and Bund tunnel layout.

As noted in the introduction, there are various methods to predict the settlement induced by tunnel construction. The Peck formula has been widely used because of its easy calculation and reasonable result.

The settlement S is evaluated by

$$S = S_{max} \exp\left(\frac{-x^2}{2i^2}\right) \tag{4.1}$$

where S_{max} is the maximum settlement that occurs above the tunnel axis, x represents the horizontal distance from the calculation point to the tunnel axis, and i is the distance from the point of inflexion of the settlement trough to the tunnel centerline. The maximum settlement is determined by the volume loss of soil V_{loss} and the width of the settlement trough i. It can be evaluated by Eq. (4.2):

$$S_{max} = \frac{V_{loss}}{\sqrt{2\pi i}} = \frac{0.313 \cdot V_l D^2}{i} \tag{4.2}$$

where V_l is the volume loss ratio of the soil, and is usually taken as approximately 0.5%–0.6% in Shanghai from past experience; D is the diameter of the tunnel.

The dimension i is the approximate function of the tunnel depth z_0, and is expressed as

$$i = K_{z0} \tag{4.3}$$

where K is the width parameter of the settlement trough; its value varies with soil layers (Sugiyama et al., 1999). Considering the properties of soil in Shanghai and past experience, it can be taken at 0.5; z_0 is the tunnel depth, the vertical distance from the surface to the tunnel centerline. For the tunnel segment adjacent to the Pujiang Hotel, the $z0$ is approximately 23.8 m.

The settlement of the Pujiang Hotel was evaluated using an empirical formula, and the results are depicted in Fig. 4.6. The maximum settlement is approximately 34 mm, the differential settlement of the foundation is about 25 mm, and the increment of the local inclination of the foundation is approximately 0.18%. The alarm value of the settlement and inclination given by the detection reports were set at 20 mm and 0.1%, respectively.

Strictly speaking, the Peck formula is only applicable for green field installations. To investigate the influence on the hotel more accurately, a 3D finite element method (FEM) simulation was used before construction. The whole analytical model consisted of the foundation of the Pujiang Hotel, and the segment of the tunnel and the surrounding soil. The geometry size of the 3D model was 90 m × 50 m × 50 m. The foundation of the hotel was modeled by an elastic model, the segment of the tunnel was modeled by an elastoplastic model, and the surrounding soil was simulated using the Mohr–Coulomb criterion.

The results show that the foundation of the hotel near the tunnel has a settlement of 25 mm and the settlement of the foundation away from the tunnel is 2 mm; the differential settlement of the foundation reaches 23 mm, with the local inclination of the foundation being 0.17%. As a result of the tunneling, a 6.5 mm horizontal displacement of the foundation occurred.

Fig. 4.87 shows that the predicted settlement curve from the Peck formula and the FEM are similar in shape, but the settlements calculated by the Peck formula are larger than that of the FEM on the whole.

2. *Deformation analysis of the Pujiang Hotel with the bored piles*

As the predicted settlement and local inclination of the foundation was expected to exceed the limit value as a result of the tunneling, it was necessary to take mitigation measures to protect the hotel. Considering that the minimum horizontal distance from the foundation to the tunnel side was only 1.7 m, the protection scheme using bored piles was proposed to separate the foundation of the hotel from

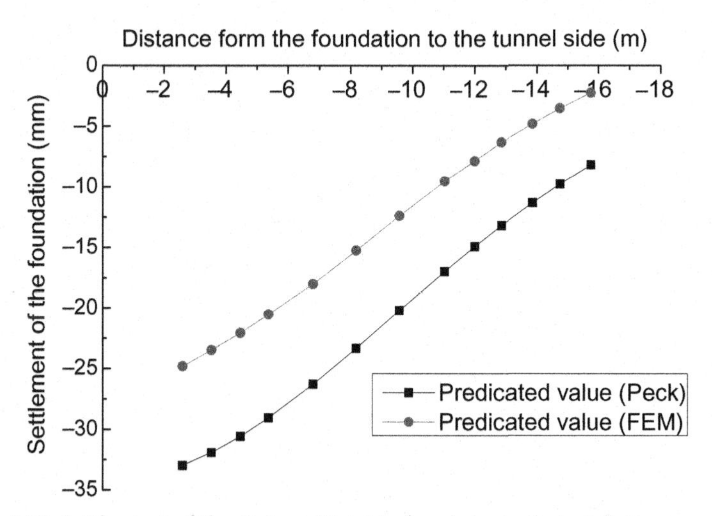

Figure 4.87 Settlement of the Pujiang Hotel before bored pile installation.

the area influenced by the tunneling. As a result, 800 mm diameter bored piles with a pile length of 32–36 m were adopted, with the length of the isolation measures being approximately 103 m and the interpile soil being reinforced using grouting. The protective effect was analyzed with numerical simulation.

After the bored piles construction, a maximum settlement of 6 mm occurred at the foundation near the tunnel, and a heave of 1 mm occurred at the other end. The differential settlement decreased from 23 to 7 mm. The settlements of the foundation before and after the pile installation are shown in Fig. 4.88. The results show that bored piles significantly reduced the effects on the Pujiang Hotel during the shield tunneling.

3. *Field monitoring of the Pujiang Hotel*

Before the shield passed through the Pujiang Hotel, 84 bored piles were constructed as an underground isolation wall. The underground isolation wall has a length of 103 m. The depth of the underground isolation wall varies from 36 to 32 m in accordance with the distance between the shield tunnel and the underground isolation wall. The deformation of the Pujiang Hotel and adjacent buildings due to the bored pile installation and the shield tunneling were both monitored, and the layout of monitoring points is shown in Fig. 4.89.

The settlement data of the monitoring points F1–F9 during the bored pile installation is presented in Fig. 4.89. On September 1, 2008,

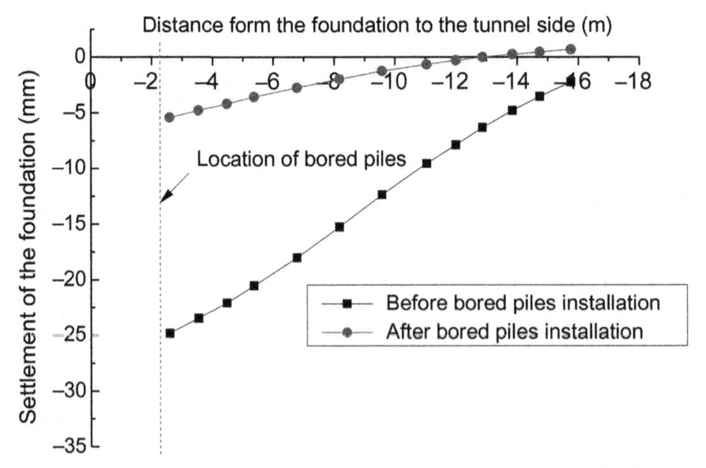

Figure 4.88 Settlement of the Pujiang Hotel foundation before and after the bored pile installation.

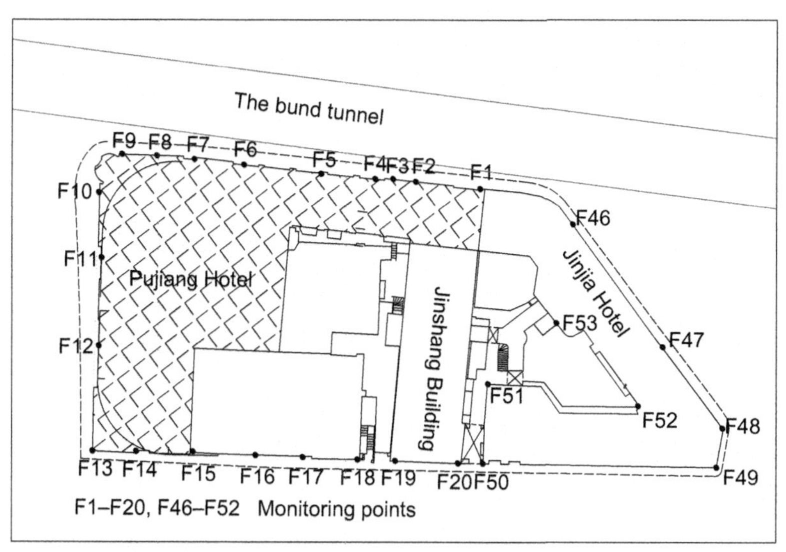

Figure 4.89 Layout of the settlement monitoring points.

13 bored piles were installed in the area corresponding to the monitoring points F5–F8. The maximum settlement occurred at F7, and the differential settlement was approximately 7 mm. Another 37 bored piles were used to isolate the area from F3 to F9 and were finished on October 10, 2008. A maximum settlement of 12 mm was observed at F6 and the minimum settlement of 0.5 mm occurred at F1; the differential settlement was 11.5 mm (Fig. 4.90). The remaining 23 bored piles along the area corresponding to the monitoring points F1–F9 were constructed from October 10, 2008, to November 13, 2008. The monitoring point F1 had a maximum settlement of 16 mm and the monitoring point F9 had a minimum settlement of 6.2 mm, and thus the differential settlement was 9.8 mm. When all 84 bored piles were completed on December 9, 2008, the maximum settlement was 12.4 mm at F3 and the minimum settlement was 6.1 mm at F8; the differential settlement was thus 6.3 mm, and the 20 mm settlement limit was not exceeded.

From monitoring data, we can see that the settlement close to the bored piles was more significant than in other regions. The main reason for this phenomena is the ground loss caused by the bored pile installation. The average settlement of the foundation increased with the increase of the bored piles, whereas the differential settlement decreased. Furthermore, in order to control the settlement more

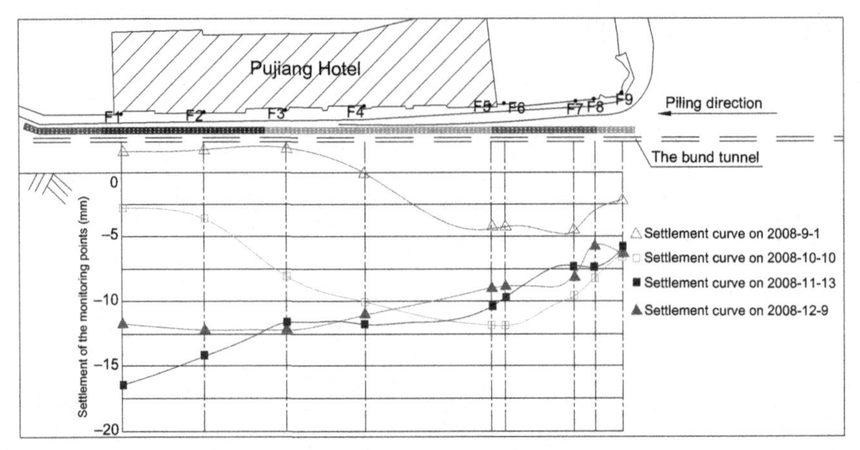

Figure 4.90 Settlement variation of the Pujiang Hotel due to the bored pile installation.

effectively, the compaction grouting was adopted during the bored pile construction process. The settlements at F1 and F2 on December 9, 2008, were less than those on November 13, 2008, due to the hysteresis effect of the compaction grouting.

Fig. 4.91A shows the settlement of the monitoring points F1, F5, and F7. The grouting ratio λ represents the ratio of the grouting volume to the theoretical value of shield tail void. The variation cure of the grouting ratio is presented in Fig. 4.91B. The results indicated that the settlement of the Pujiang Hotel had no obvious change during the passage of the shield, whereas the change of the grouting ratio led to variation of the settlement, with the settlement increasing with the decrease of the grouting ratio. The settlement of the Pujiang Hotel is illustrated in Fig. 4.90 when the shield tail was 90 m away from the hotel. The monitoring data show that the maximum settlement among the monitoring points was 9.8 mm (F47), whereas the maximum settlement of the Pujiang Hotel was 6.7 mm, which was close to the numerical result of 6 mm. The differential settlement of the Pujiang Hotel was 6.4 mm, which agreed well with the FEM calculated value of 7 mm (Fig. 4.92).

The deformation of the monitoring points in the protected area can be divided into two stages: the first was the bored pile installation stage. The deformation of the first stage was caused by the soil disturbance and the loss of soil due to the bored pile installation. The second was the tunneling stage. The settlement was induced by tunneling. As the

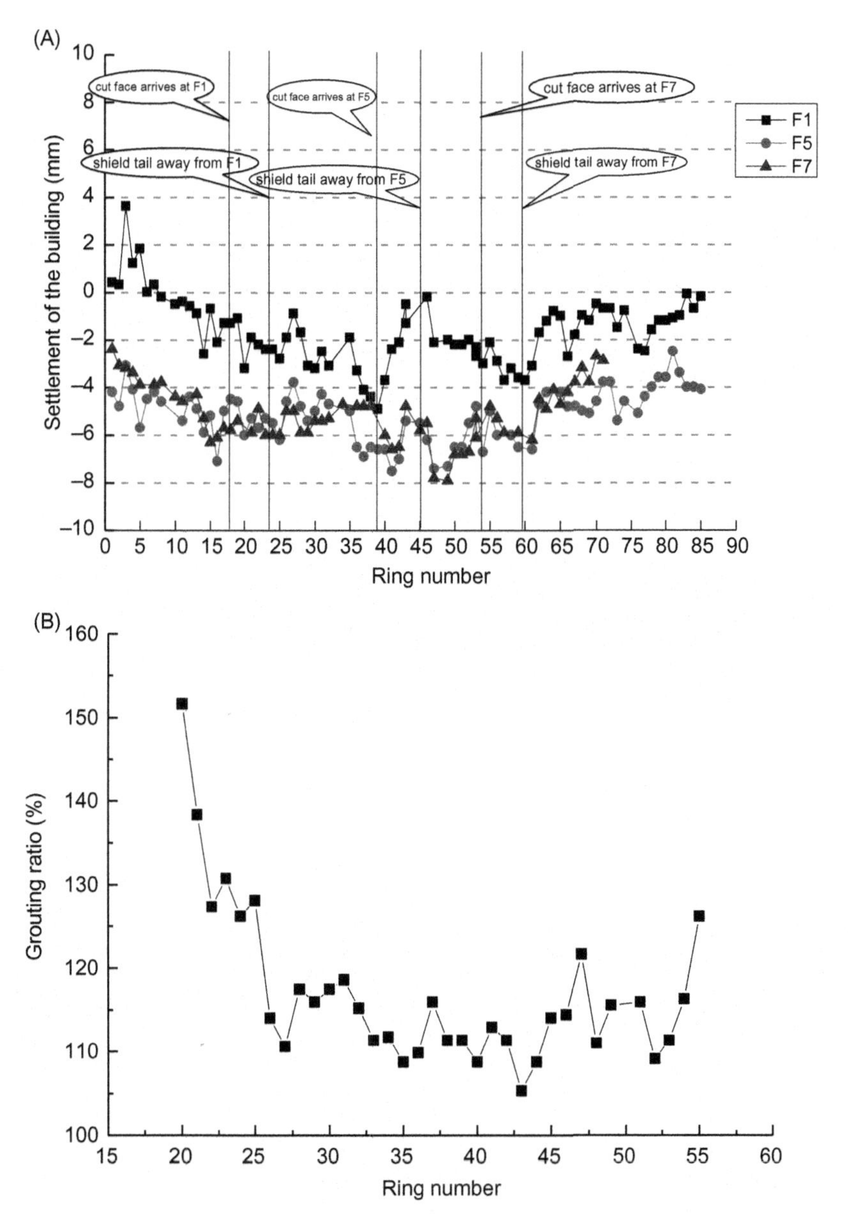

Figure 4.91 (A) Settlement curve of the Pujiang Hotel foundation. (B) Grouting ratio along the tunnel.

settlement from the construction process of bored pile was much smaller than the settlement resulting from the tunneling, using the isolation wall built by bored piles was effective.

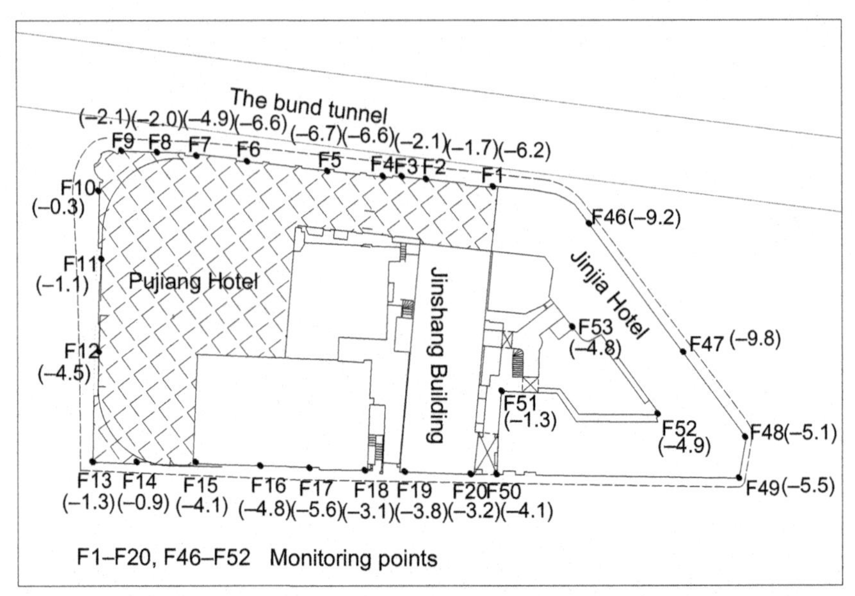

Figure 4.92 Settlement of the monitoring points when the tail is 90 m away from the Pujiang Hotel. The unit of deformation value given in the brackets is mm.

Although the monitoring points F1−F9 are closer to the tunnel than the monitoring points F46−F47, their settlements were less than that of F46−F47. The main reason was because the monitoring points F1−F9 were located in the protected area of the bored piles. The field monitoring results indicated that the bored piles were an effective mitigation measure.

4.5.3.2 Analysis of Influence on the Shanghai Mansion

The Shanghai Mansion is on the other side of the Bund tunnel adjacent corresponding to the Pujiang Hotel, and was built in 1934. It is an early example of modern architecture. The main building uses a steel frame structure and the foundation a piled raft. The attached building uses a composite foundation of pile and strip. The horizontal distance from the building to the tunnel was within the range of 5.2−5.9 m. The separation between the Shanghai Mansion and the Bund Tunnel is shown in Fig. 4.93. None significant differential settlement of the building was recorded before tunnel construction. Considering the structure situation of the building, its distance to the tunnel, the cost, and the construction duration, the grouting method was proposed to reduce the effects of tunneling on the Shanghai Mansion.

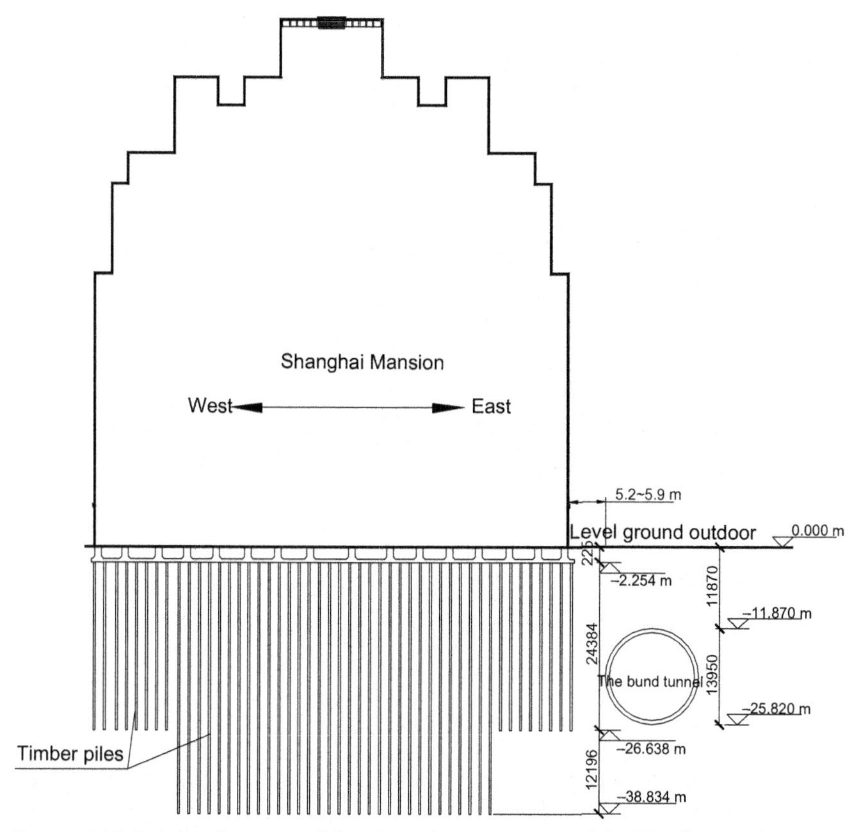

Figure 4.93 Relative location of the Shanghai Mansion and the Bund Tunnel.

The reinforcement effect of the grouting was verified with a field test. Fig. 4.94 shows the layout of the monitoring points. The monitoring data collected is presented in Table 4.5 when the tail passed the Shanghai Mansion; it was approximately 90 m away from the building. Positive and negative values indicate heave and settlement of ground, respectively. The results show that the primary displacement of the Shanghai Mansion was heave, a maximum uplift of 5.3 mm was recorded at F21, and the minimum settlement was 3.2 mm, observed at F25. The differential settlement was thus 8.5 mm. As can be seen, the grouting method was also effective at protecting the building close to the tunnel during construction.

A comparative analysis of displacement of the Pujiang Hotel and Shanghai Mansion was conducted. Fig. 4.95 plots the displacement of the two buildings: the positive displacement represented the heave and the negative displacement represented the settlement. The monitoring data

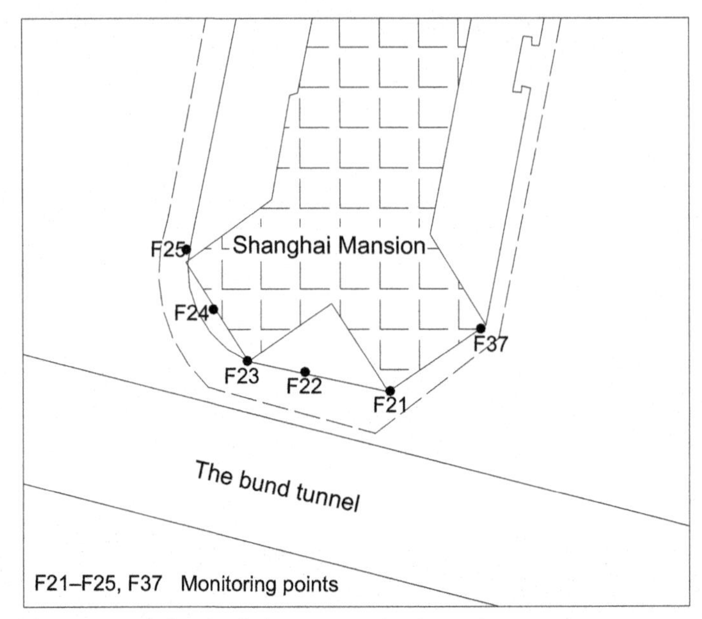

Figure 4.94 Layout of the displacement monitoring points.

Table 4.5 Displacement of the Shanghai Mansion

Monitoring point	Vertical displacement (mm)	Monitoring point	Vertical displacement (mm)
F21	5.3	F24	3.7
F22	4.3	F25	− 3.2
F23	4.8	F37	2.4

show that the deformation of the Shanghai Mansion was more significant than that of the Pujiang Hotel, and the main reason for this was that the grouting method could not fully isolate the influence of the construction. Therefore, the isolation wall with bored piles had a better protective effect than the grouting method.

4.5.3.3 Analysis of Influence on the Other Buildings Along the Bund

As noted earlier, there were many other historical buildings with different architectural styles adjacent to the Bund tunnel. It was essential to take mitigation measures to protect these buildings and reduce the tunneling influence on them. Considering that the minimum horizontal distance from the buildings to the external boundary of the Bund tunnel was within the range of 9.9–30 m, the main mitigation adopted was to carefully control

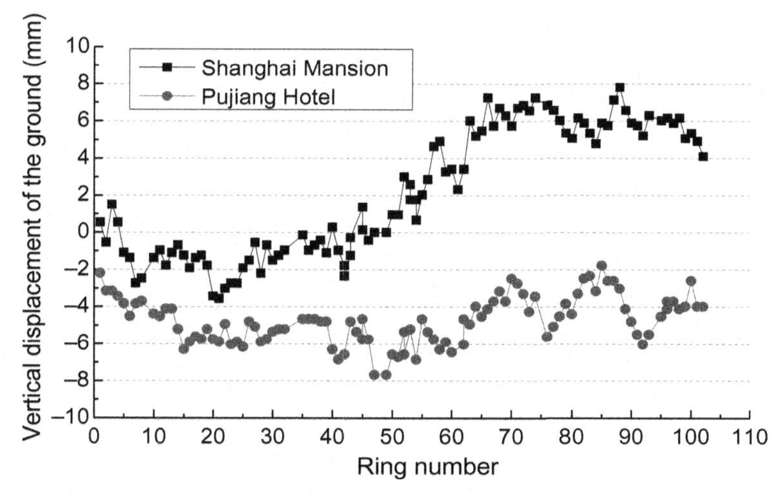

Figure 4.95 Comparative analysis of the settlement of the Pujiang Hotel and Shanghai mansion.

the driving parameters including the soil chamber pressure, the tunneling velocity, the shield posture, and the quantity and pressure of the synchronized grouting. The ratio of the soil chamber pressure to the vertical earth pressure at the center of the excavation face was in the range of 0.75–0.87. The tunneling velocity was approximately 25 mm/min during the normal tunneling process, and there was a15 mm/min deviation during the steering correction process. The grouting volume was 17–20 m^3 for each ring. The following selected buildings further reveal the protection result with the abovementioned mitigation.

1. *Field monitoring of the Youbang Mansion*

   The Youbang Mansion was built in the early 20th century, in the renaissance architecture style. The initial inclination ratio to the north of the building was in the range of 3.7%–7.4%, and the slopes westward had an inclination ratio of 0.1%–5.5%. The layout of the monitoring points and the settlement of the building during the construction are shown in Fig. 4.96. The horizontal distance from the monitoring points J1 and J2 to the tunnel axis were 22.0 and 22.6 m, respectively, and the vertical distance from the monitoring points to the tunnel center was 15.1 m. The predicted value of the settlement points J1 and J2 using the Peck empirical formula were 0.8 and 0.6 mm. The monitoring deformation was within ± 1 mm during the shield driving process in general, and the differential settlement of the building was 1 mm. Both the prediction of the settlement and the on-site observation value of the Youbang mansion were less than 1 mm. On the other hand, the

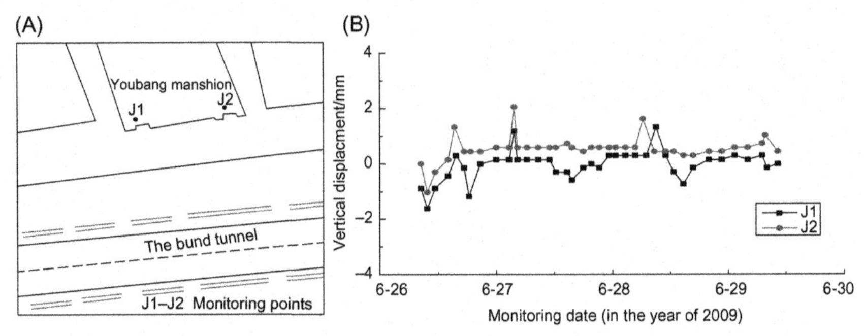

Figure 4.96 The (A) layout and (B) settlement of the Youbang Mansion.

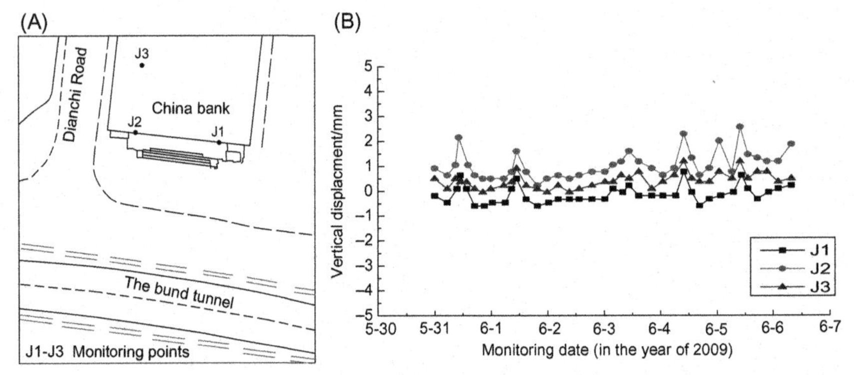

Figure 4.97 Layout and displacement of the Bank of China building.

monitoring data indicated that the ground was slightly uplifted at times during the tunneling process due to the synchronized grouting.

2. *Field monitoring of the Bank of China building*

The Bank of China building was designed as traditional Chinese architecture; it is a steel frame structure that has been in service for more than 70 years. Structural deterioration including cracking, corrosion, and leakage was observed; the structure also inclines to the northwest and has significant local differential settlement. Fig. 4.97 shows the layout of the monitoring points and the displacement of the building. The horizontal distance from the monitoring points was in the range of 29–39.2 m, and the predicted settlement using the Peck formula was within 0.1 mm. As can be seen, there was hardly any tunneling effect on the China Bank building. The on-site monitoring results also showed that there was no obvious deformation during the shield tunneling, and the deformation of the monitoring points J1, J2, and J3

Table 4.6 displacement of the building group along the Bund

Building name	Minimum horizontal distance to the tunnel center (m)	Vertical distance to the tunnel center (m)	Final vertical displacement (mm)	Prediction deformation with Peck formula (mm)
China Ever bright Bank	25.9	25.2	− 1.2	− 3.7
Culture Film and TV Corporation	26.4	23.1	− 2.2	− 2.4
Foreign Trade office	32.0	21.2	0.9	− 0.3
Agricultural Bank of China	21.0	19.4	− 3.4	− 3.8
Industrial and Commercial Bank of China	28.0	17.7	2.6	− 0.3
Bank of China	29.0	16.6	− 0.5	− 0.1
North Building of the Peace Hotel	28.7	15.3	3.7	− 0.05
South Building of the Peace Hotel	29.8	15.3	2.5	− 0.03
CITIC Bank	37.0	15.1	− 0.7	− 0.0
Textiles Import and Export Corporation	27.0	15.0	− 0.6	− 0.1
Youbang Mansion	22.0	15.1	− 2.3	− 0.8
China Merchants Bank	24.0	15.0	− 1.8	− 0.4
Foreign Exchange Trading Centre	31.0	15.1	2.0	− 0.01
Bund Law Office	27.0	15.1	− 0.6	− 0.1
General Labour Union	16.9	15.0	− 5.2	− 4.8
Shanghai Customs	18.0	15.0	− 3.1	− 3.4
Shanghai Pudong Development Bank	20.0	15.1	− 3.1	− 2.0

displayed the same tendency, with J1 being settlement, J2 being heave, and J3 exhibiting the smallest deformation of within 0.6 mm, with the differential settlement being 1 mm.

3. *Field monitoring result of the building group*

During the tunneling period, the deformations of the other buildings along the tunnel were also monitored, and the final displacement of the building group is presented in Table 4.6. Positive and negative values indicate the heave and settlement of the ground, respectively.

As shown in Table 4.6, the observed settlement of most buildings was less than the empirically predicted settlement, and the value of the observed settlement varied with the tunnel depth and the horizontal distance. The tunneling construction had the most influence on the Labour Union building, with a building settlement of 5.2 mm. The main reason for this is that this building is the closest to the tunnel among all those listed in Table 4.6. As deformations of all buildings do not exceed the permissible value, the building group was effectively protected by careful selection of shield driving parameters.

4.6 CONCLUSIONS

1. Three different types of mitigation methods including bored piles, grouting, and tunneling parameter optimization were adopted to protect the buildings along the tunnel according to the relative location of the buildings and the tunnel.
2. The deformation monitoring data of the Pujiang Hotel showed that the isolation wall with bored piles with grouting had a very good protective effect on buildings within 5 m away from the tunnel. The deformation of the Shanghai Mansion shows that grouting method was effective for those buildings, which were 5 to 10 m away from the tunnel. the range of 5–10 m. Optimization of the tunneling parameters reduced the influence on the buildings that were more than 10 m away from the tunnel.
3. Comparative analysis of the deformation of the Pujiang Hotel and Shanghai Mansion illustrates that the isolation wall made of bored piles is more effective than grouting in Shanghai soil.

4.7 QUESTIONS

4.1. It is known that the drill-and-blast method causes serious ground vibrations. How can these vibrations be reduced or prevented?

4.2. Choose the suitable TBM types for these ground conditions (a. Gripper TBM; b. EPB TBM; c. Single Shield TBM; d. Slurry Pressure Balance TBM; e. Double Shield TBM; f. Pipe Jacking).

- Soil ( )
- Rock ( )

4.3. When a tunnel is excavated, ground settlement is inevitable. Choose the right relationship between the following four values and explain why.

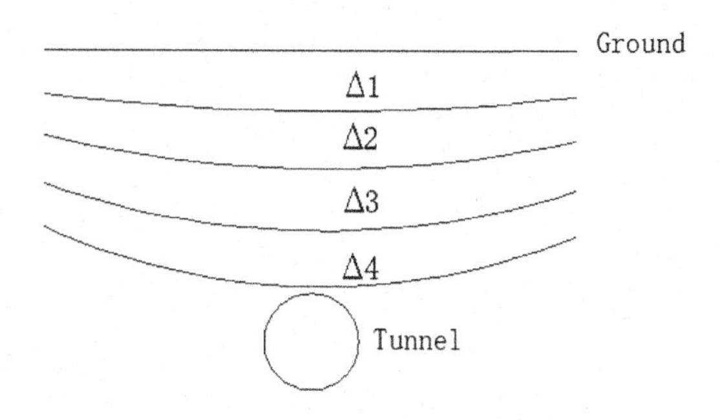

a. $\Delta_1 > \Delta_2 > \Delta_3 > \Delta_4$
b. $\Delta_1 > \Delta_2 = \Delta_3 > \Delta_4$
c. $\Delta_1 = \Delta_2 = \Delta_3 = \Delta_4$
d. $\Delta_4 > \Delta_3 = \Delta_2 > \Delta_1$
e. $\Delta_4 > \Delta_3 > \Delta_2 > \Delta_1$

4.4. Describe the application scope of immersed tube tunneling.

4.5. What does RQD and RMR stand for, and how are they used?

4.6. What is the construction sequence of conventional tunneling?

4.7. State the general construction sequence of the cut-and-cover method in soft ground formation.

4.8. As shown in the following figure, why do people use slurry shield (or mix shield) more often in very big diameter tunnels?

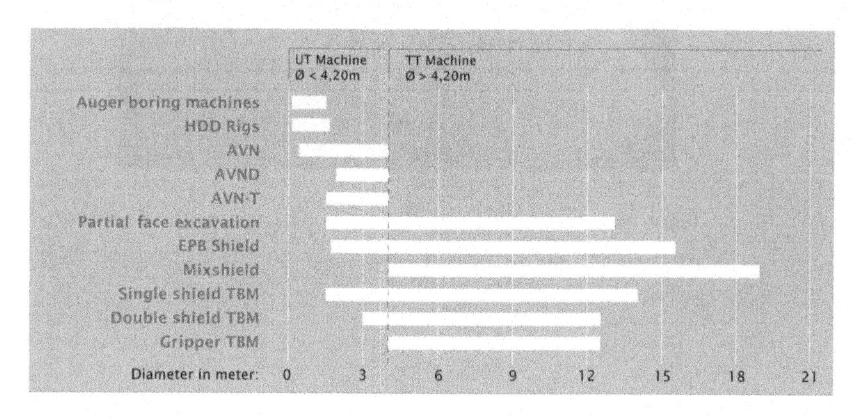

4.9. What are the advantages and disadvantages of shield tunneling?

4.10. What are the main contributors to the surface settlement trough in TBM tunneling?

4.11. Which of the following is most efficient in terms of excavation speed and why? Explain which plan is safest in terms of stability and why.

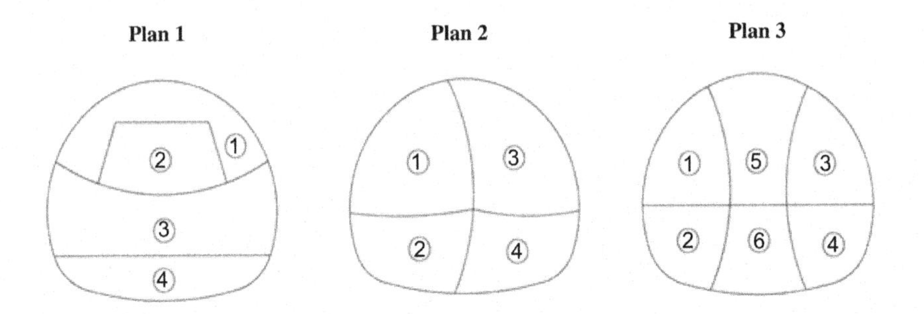

4.12. For the excavation sequence in the image below, why is the rock in the invert part excavated first and the rock in the core part excavated last?

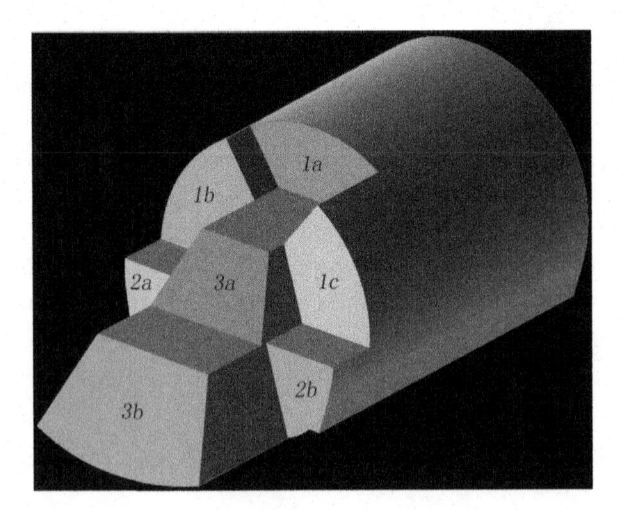

4.13. If you want to build a link between two places separated by a water body (strait, river, etc.), what construction methods are best? Why? What are the major challenges?

4.14. Explain the advantages and disadvantages of FDM, FEM, and BEM.

4.15. What are the main differences between pipe jacking and TBM?

4.16. Another name proposed for the submerged floating tunnel (see figure below) is Archimedes bridge. Which one do you think is better? Why? What are the main challenges of this kind of structure?

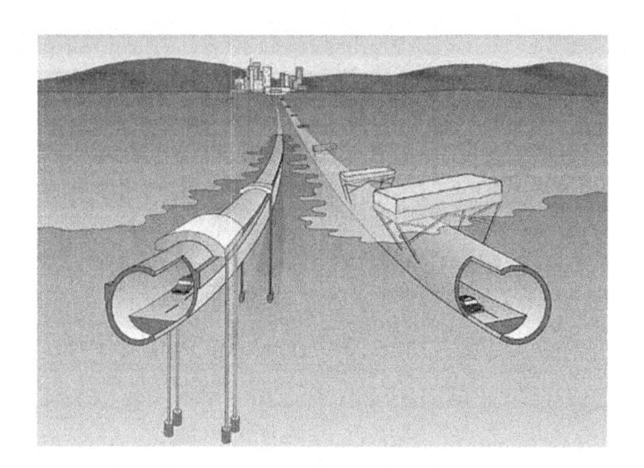

REFERENCES

Attewell, P. B., & Woodman, J. P. (1982). Predicting the dynamics of ground settlement and its derivatives by tunneling in soil. *Ground Eng.*, *15*, 13–22.

Bauer. (2012). *Soil mixing wall—system*. Retrieved from <http://www.bauerpileco.com/export/shared/documents/pdf/bma/datenblatter/info_39_e.pdf>.

Bickel, J. O., Kuesel, T. R., & King, E. H. (1996). *Tunnel engineering handbook* (2nd ed.). New York: Chapman & Hall.

Chakeri, H., Ozcelik, Y., & Unver, B. (2013). Effects of important factors on surface settlement prediction for metro tunnel excavated by EPB. *Tunn. Undergr. Space Technol.*, *36*, 14–23.

Chen, L. T., Poulos, H. G., & Loganathan, N. (1999). Piles response caused by tunneling. *J. Geotech. Geoenviron. Eng.*, *125*, 207–215.

Chou, W. I., & Bobet, A. (2002). Predictions of ground deformations in shallow tunnels in clay. *Tunn. Undergr. Space Technol.*, *17*, 3–19.

Durabond. (n.d.). *Layout of slurry pipe jacking system*. Retrieved from <http://www.durabond.com.tw/en/product1078.html?eproducts_cat_id = 103&eproducts_id = 131>.

Dywidag-Systems International. (2014). *The Mandacarú tunnel in Bahia: Dywidag rock bolt secure Brazil's future as an exporter*. Retrieved from <https://www.dsiunderground.com/projects/project-details/article/the-mandacaru-tunnel-brazil.html>.

Facesupport.org. (n.d.). *Face support pressure and its importance*. Retrieved from <http://www.facesupport.org/wiki>.

Fargnoli, V., Boldini, D., & Amorosi, A. (2013). TBM tunneling-induced settlements in coarse-grained soils: the case of the new Milan underground line 5. *Tunn. Undergr. Space Technol.*, *38*, 336–347.

Farrell, R.P., 2010. Tunneling in sands and the response of buildings. Ph.D Thesis, University of Cambridge, UK.

Franki Foundations. *Business units and products, soil mix: Mix-in place wall.* Retrieved from <http://www.ffgb.be/Business-Units/Retaining-Walls---Utilities/Mixed-in-place-wand.aspx?lang = en-US>.

González, C., & Sagaseta, C. (2001). Patterns of soil deformations around tunnels. Application to the extension of Madrid Metro. *Comput. Geotech.*, *28*, 445–468.

Harris, D. I., Mair, R. J., Love, J. P., et al. (1994). Observations of ground and structure movements for compensation grouting during tunnel construction at Waterloo station. *Geotechnique*, *44*, 691–713.

Herrenknecht. (n.d.b). *Partial-face excavation machine—Economical tunnelling technology with high flexibility.* Retrieved from <https://www.herrenknecht.com/en/products/core-products/tunnelling/partial-face-excavation-machine.html>.

Hisatake, M. (2011). A proposed methodology for analysis of ground settlements caused by tunneling, with particular reference to the "buoyancy" effect. *Tunn. Undergr. Space Technol.*, *26*, 130–138.

Höfler, J., Schlumpf, J., & Jahn, M. (2011). *Sika sprayed concrete handbook.* Sika Services AG & Putzmeister AG, Zürich, Switzerland.

ITA Working Group n°11. (2016a). *ITA report n°7—An owners guide to immersed tunnels.*

ITA Working Group n°11. (2016b). *ITA report n°7—An owners guide to immersed tunnels—annexes.*

Lee, A.H.S. (2007). Engineering survey system for TBM (tunnel boring machine) tunnel construction. In *FIG working week.* Hong Kong.

Lee, C. J. (2012). Three-dimensional numerical analyses of the response of a single pile and pile groups to tunneling in weak weathered rock. *Tunn. Undergr. Space Technol.*, *32*, 132–142.

Lee, K. M., & Rowe, R. K. (1990a). Finite element modeling of the three-dimensional ground deformations due to tunneling in son cohesive soils: Part I-method of analysis. *Comput. Geotech.*, *10*, 87–109.

Lee, K. M., & Rowe, R. K. (1990b). Finite element modeling of the three-dimensional ground deformations due to tunneling in son cohesive soils: Part II-results. *Comput. Geotech.*, *10*, 111–138.

Loganathan, N., & Poulos, H. G. (1998). Analytical prediction for tunneling-induced ground movement in clays. Geotech. *Geoenviron. Eng*, *124*, 846–856.

Lunniss, R., & Baber, J. (2013). *Immersed tunnels.* Boca Raton, FL: CRC Press.

Maidl, B., Herrenknecht, M., Maidl, U., & Wehrmeyer, G. (2012). *Mechanised shield tunnelling* (2nd ed). Berlin, Germany: Wihelm Ernst & Sohn.

Mining & Construction. (2013). *The Scaletec effect at Tara mines.* Retrieved from <https://miningandconstruction.com/cat/undergroundmining/scaling/>.

NFM Technologies. (n.d.). *Slurry TBM (Benton' Air®).* Retrieved from <http://www.nfm-technologies.com/-Soft-ground-machines-.html>.

O'Reilly, M. P., & New, B. M. (1982). *Settlements above tunnels in the United Kingdomtheir magnitude and prediction, . Tunnellling* (vol. 82, pp. 173–181). London: IMM.

Oh, T.M., Joo, G.W., Hong, C.H., Cho, G.C., & Ji, I.T. (2013). Tunnel excavation using waterjet pre-cutting technology. In *World tunnel congress 2013 Geneva Underground—The Way to the Future*, pp. 1567–1570.

Ou, C.-Y. (2006). *Deep excavation, theory and practice.* Taipei, Taiwan: National Taiwan University of Science and Technology.

Pang, C.H., Yong, K.Y., Chow, Y.K., 2005. Three-dimensional numerical simulation of tunnel advancement on adjacent pile foundation. In: Proceedings, ITA-AITES 2005 World Tunnel Congress, pp. 1141–1148.

Pearson Scott Foresman. (2009). *Line art drawing of a dredge* [Drawing]. Retrieved from <https://commons.wikimedia.org/wiki/File:Dredge_(PSF).png>.

Peck, R.B., 1969. Deep excavations and tunneling in soft ground. In: Proc. 7th Int. Conf. Soil Mech. & Found. Eng., Mexico, State of the Art Volume, pp. 225–290.

Pipe Jacking Association. (2017). *An introduction to pipe jacking and microtunnelling.* Retrieved from <http://www.pipejacking.org/assets/pj/static/PJA_intro.pdf>.

Rowe, R. K., & Kack, G. J. (1983). A method of estimating surface settlement above tunnels constructed in soft ground. *J. Geotech. Can.*, *20*, 11–22.

Sagaseta, C. (1987). Analysis of undrained soil deformation. *Geotechnique*, *37*, 301–320.

Sugiyama, T., Hagiwara, T., Nomoto, T., et al. (1999). Observations of ground movements during tunnel construction by slurry shield method at the docklands light. *Soils Found.*, *39*(3), 99–112.

The National Standards Compilation Group of People's Republic of China. (1994). *GB 50218–94 standard for engineering classification of rock masses.* Beijing: China Planning Press. (in Chinese).

Toraño, J., Rodríguez, R., Diego, I., & Rivas, J. M. (2006). Estimation of settlements due to shallow tunnels and their effects. *Tunn. Undergr. Space Technol.*, *21*, 288–294.

Trelleborg. (2009). *Omega seals.* Product brochure retrieved from <http://www.trelleborg.com/en/engineered-products/products--and--solutions/tunnel--seals/omega--seals>.

Trelleborg. (n.d.). *Gina gasket.* Product brochure retrieved from <http://www.trelleborg.com/en/engineered-products/products--and--solutions/tunnel--seals/gina--seals>.

Tunnel Ausbau Technik. (n.d.). *Template for shotcrete application and Profile for machine excavation.* Retrieved from <http://www.tat-befer.de/pante_en.htm>.

US Department of Transportation Federal Highway Administration. (2009). *Technical manual for design and construction of road tunnels—Civil elements.* Publication no. FHWA-NHI-10-034.

US Department of Transportation Federal Highway Administration. (2015). *Soil nail walls—Reference manual.* Publication no. FHWA-NHI-14-0.

Verruijt, A., & Booker, J. R. (1996). Surface settlements due to deformation of a tunnel in an elastic half plane. *Geotechnique*, *46*, 753–756.

Wisser, C., Augard, C. E., & Burd, H. J. (2005). Numerical modelling of compensation grouting above shallow tunnels. *Int. J. Numer. Anal. Methods Geomech.*, *29*, 443–471.

Xiang, Y. Y., Jiang, Z. P., & He, H. J. (2008). Assessment and control of metro-construction induced settlement of a pile-supported urban overpass. *Tunn. Undergr. Space Technol.*, *23*, 300–307.

Xu, K. J., & Poulos, H. G. (2001). 3-D elastic analysis of vertical piles subjected to "passive" loadings. *Comput. Geotech.*, *28*, 349–375.

Xu, Q. W., Zhu, H. H., Ding, W. Q., & Ge, X. R. (2011). Laboratory model tests and field investigations of EPB shield machine tunneling in soft ground in Shanghai. *Tunn. Undergr. Space Technol.*, *26*, 1–14.

FURTHER READING

Carroll, C. (2008). *Subsea connection for desalination plant.* Retrieved from <https://www.tunneltalk.com/Subsea-connection-for-desalination-plant.php>.

China Petroleum Pipeline Engineering. (2011). *The shield of the west second line successfully completed sinking* [photo]. Retrieved from <http://guandaoxinsijun.blog.sohu.com/175326538.html> (in Chinese).

Ching, F. D., Faia, R. S., & Winkel, P. (2006). *Building codes illustrated: A guide to understanding the 2006 international building code* (2nd ed.). New York: Wiley.

EFSCCS—European Federation of Specialist Construction Chemicals and Concrete Systems. (2008). *Specification and guidelines on thin spray-on liners for mining and tunneling.* Retrieved from <http://www.efnarc.org/pdf/ENC%20250TSL%20v7.2%2025-07-08_r1.pdf>.

Herrenknecht. (2009). *Herrenknecht cutting tools—Highest quality claims in china.* Retrieved from <https://www.herrenknecht.com/de/produkte/zusatzequipment/vortrieb-logistik/abbauwerkzeuge.html?tx_torrmediabasket_mediabasket%5Bpath%5D = %2Fuploads%2Fmedia%2FChina_Cutting_Tools_Folder_GB_101209_01.pdf&tx_torrmediabasket_mediabasket%5Baction%5D = downloadSinglePDF&tx_torrmediabasket_mediabasket%5Bcontroller%5D = Download&cHash = d0e38f7436dcb1ea503bb37024ae7082>

Herrenknecht. (n.d.a). *EPB shield—Fast tunnelling technology with a broad application range.* Retrieved from <https://www.herrenknecht.com/en/products/core-products/tunnelling/epb-shield.html>.

Herrenknecht. (n.d.c). *Gripper TBM—Experts for tough hard rock.* Retrieved from <https://www.herrenknecht.com/en/products/core-products/tunnelling/gripper-tbm.html>.

Herrenknecht. (n.d.d). *Mixshield—Safe tunnelling technology for heterogeneous ground.* Retrieved from <https://www.herrenknecht.com/en/products/core-products/tunnelling/mixshield.html>.

ITA. (n.d.). *Subaquatic tunneling.* Retrieved from <http://tunnel.ita-aites.org/en/how-to-go-undergound/construction-methods/subaquatic-tunnelling>.

OTN. (n.d.). *Diving supervision when immersing tunnel elements.* Retrieved from <http://www.otnbv.com/specialisms/duikbegeleiding-afzinken-tunnelelementen/#>.

Pents, V. (2012). *World-class tunnel lining.* Retrieved from <http://civilpents.blogspot.nl/2012/06/world-class-tunnel-lining.html?m = 1>.

Peri. (n.d.). *Øresund Link, Denmark.* Retrieved from <https://www.peri.com/en/projects/civil-engineering/oeresund-link.html>.

Railway Gazette. (2013). *First crossrail tunnel completed.* Retrieved from <http://www.railwaygazette.com/news/single-view/view/first-crossrail-tunnel-completed.html>.

Robbins. (n.d.). *Double shield tunnel boring machine (TBM).* Retrieved from <http://www.canadianundergroundinfrastructure.com/product/2575/double-shield>.

Shield Tunneling Association of Japan. (n.d.). *DOT tunneling method—Contributing to effective use of a wide range of underground spaces.* Retrieved from <http://www.shield-method.gr.jp/eng/pdf/brochure/dot.pdf>.

Tana, Y., Zhua, H., Penga, F., Karlsrudb, K., & Weic, B. (2017). Characterization of semi-top-down excavation for subway station in Shanghai soft ground. *Tunnelling and Underground Space Technology, 68*, 244–261.

The Construction Civil. (n.d.). *Caissons—Types of caissons.* Retrieved from <https://www.theconstructioncivil.org/caissons-types-of-caissons/>.

Thewes, M., & Budach, C. (2009). Grouting of the annular gap in shield tunnelling—An important factor for minimisation of settlements and production performance. In *Proceedings of the conference ITA-AITES World Tunnel Congress 2009 "Safe Tunnelling for the City and Environment"*, Budapest.

Washington Department of Transportation. (n.d.). *A truck that carries 10 tunnel liner segments into the SR 99 tunnel* [Photograph]. Retrieved from <http://www.equipmentworld.com/bertha-pushes-past-halfway-point-of-tunneling-for-washingtons-sr-99/>.

CHAPTER 5

Project Management

Contents

With advances in underground engineering and greater analytical understanding of the phenomena involved, tunnels can now be built in more challenging ground conditions. Projects have thus become more complex, calling for more specialized knowledge and improved communication networks (Muir Wood, 2000). Modern construction project management is a centralized system that plans, organizes, and controls the fieldwork to meet a project's time, cost, and quality requirements (Ritz, 1994).

Irrespective of their size, all underground engineering projects are complex. While larger-scale projects are understandably complex due to

Underground Engineering
DOI: https://doi.org/10.1016/B978-0-12-812702-5.00005-0

their work scope, smaller ones can also be. New technologies are often tested on smaller projects first (before being scaled up to bigger ones). The latter are also usually on tighter time schedules and in more remote and space-restricted locations (Ritz, 1994).

All projects can be mapped using the life cycle concept shown in Fig. 5.1. It consists of several phases as detailed in Fig. 5.2. After the conceptual phase, a project is defined and carried out before being delivered. The execution phase is the main life cycle phase as it is when most of the project resources are used, costs increase rapidly, and time is limited.

5.1 STAKEHOLDERS

A construction project involves different stakeholders who play different roles and thus have different objectives. This section introduces the three main stakeholders: the client, designers, and contractors.

5.1.1 The Client

The client's role is to provide the financial resources and services for the project at a reasonable cost and to set safety standards and safety responsibilities. These aspects of the project are defined in the contract.

To ensure construction quality, the client ideally has an experienced team to organize the project. A trusted external project management team

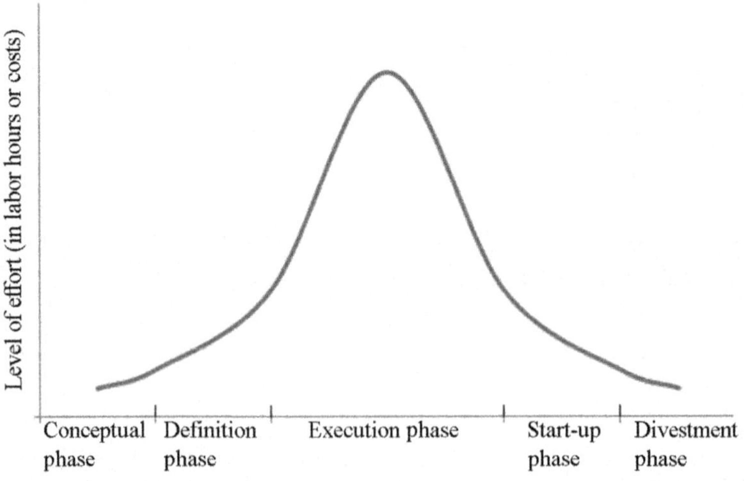

Figure 5.1 Project life cycle curve. *Adapted from Ritz, G. J. (1994).* Total construction project management *(1st ed.). New York: McGraw-Hill.*

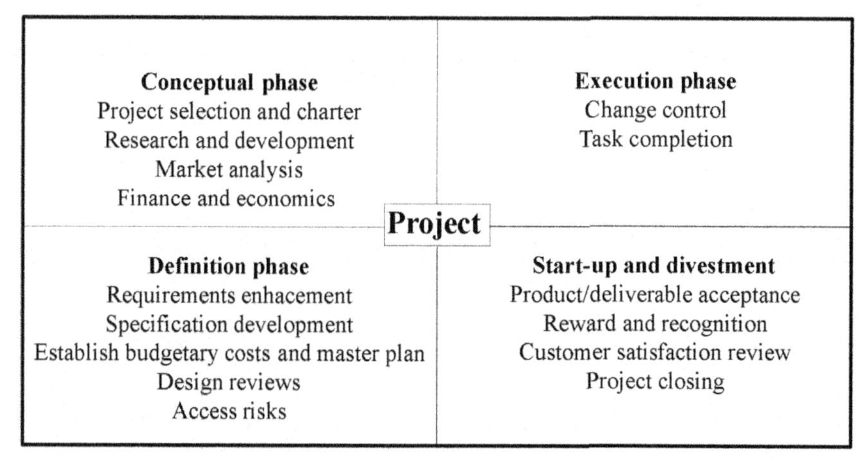

Figure 5.2 Typical activities in life cycle structure. *Adapted from Ritz, G. J. (1994).* Total construction project management *(1st ed.). New York: McGraw-Hill.*

can also be used. The role of this team is to seek suitable designers and contractors in accordance with the project goals, regarding overall project schedule, budget, quality, etc.

The bidding and contracting process is imperative for the client as this is when it chooses which companies will work on the project. During this phase, the management team selects the contractors and negotiates the contract with them. The client submits a tender consisting of an elaborate prebidding document. The contract is then established when a company is chosen. Ground rules and construction work procedures are outlined in it, in addition to technical and performance-related specifications for quality control. The overall process is summarized in Table 5.1.

5.1.2 The Designer

The designer is responsible for delivering the design documents that are requested by the client and comply with the contract. The designer can be externally hired or be part of the client company.

As few tunnel projects are alike, their design work differs between each. Since the design has a strong influence on the construction process, it is essential for this phase to be thoroughly carried out. The role of designers is therefore to strike a balance between quality, cost, and time. In order to provide a sustainable design, they must also provide low-carbon solutions.

Table 5.1 A matrix of construction project characteristics

Development phase	Contracting phase	Execution phase
Activities	Activities	Activities
Project planning	Contracting plan	Detailed engineering
Market development	Contractor screening	Procurement
Process planning	Selection of bidders	Construction
Cost estimating	Invitation for proposals	
Basic design	Contractor's proposals	
	Bid review	
	Contract award	
By client	By client and contractor	By contractor

Source: Adapted from the reproduction of Ritz, G. J. (1994). *Total construction project management* (1st ed.). New York: McGraw-Hill, from Project Management Institute seminar—symposium, 1978.

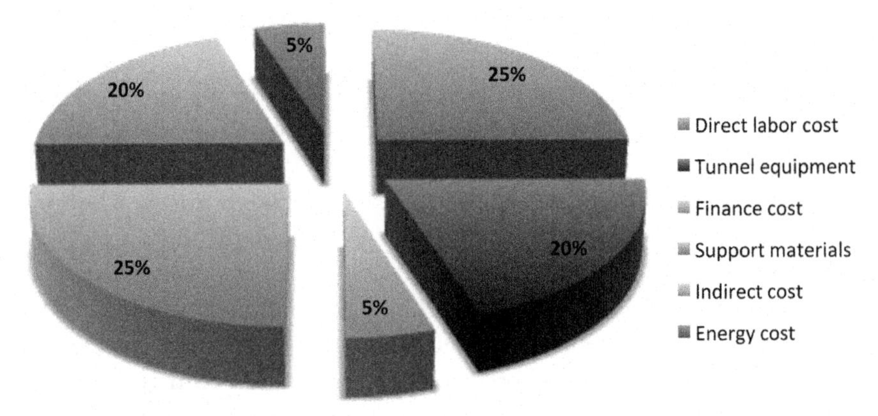

Figure 5.3 Allocation of the capital for contractors.

5.1.3 The Contractor

Once the contractor delivers the finished project to the client, the client needs to evaluate if it complies with the contract, and then accept it. Being a separate stakeholder, the contractor also aims to make a profit with each project. Contractors make bids with tender documents, in an attempt to win the contract. If successful, they must carry out the construction work as outlined in the predefined contract. Contractors must also respect the project design provided by the designer.

Fig. 5.3 shows the typical cost allocation for a tunneling project. Support materials and direct labor together make up about half of the total cost. In developing countries, direct labor costs may be lower, while equipment and supporting materials costs may be higher.

5.2 CONTRACT MANAGEMENT

Contracts are official agreements with legal value. The structure and content of a contract varies greatly from one jurisdiction to another. The two major law systems found worldwide are common law and civil law. While common law uses judicial emphasizes source, civil law places an emphasis on codes and statutes. The nature of a contract should therefore be established from the beginning of a project, in order to prevent unnecessary loss through ignorance of the law system.

5.2.1 Forms of Contracts

A construction contract contains elements such as construction objectives, stakeholder responsibilities, and modification procedures. It ensures the work is delivered to the client, payments are made, and disputes resolved. Standard contract forms exist to ensure clearly defined contractual relationships between the parties. This also improves the efficiency of the contractual process of engineering projects. The most common types of contracts are the FIDIC (International Federation of Consulting Engineers) and the NEC (New Engineering Contract).

- The FIDIC has produced internationally used standard form contracts for civil engineering projects. These contracts are distinguished by color, and aim to offer a fair allocation of risk between parties (Pinsent Masons, 2011).
- NEC is a simple language contract developed by the Institution of Civil Engineers (ICE) with the aim of introducing a new nonadversarial (Dickson, 2013) contract strategy with more effective and smoother project management (NEC, n.d.). These contracts are growing in popularity worldwide as they offer a more collaborative management philosophy (Musaka, 2016).

Companies develop suites of contracts in which there are different standard contracts. This allows them to apply for a range of construction projects, ranging from large-scale ones to individual homes. The stakeholders must have a deep understanding of the form of the contract and its specifics.

As an example, Table 5.2 gives the major differences between the FIDIC and Chinese standards.

Table 5.2 Major differences between FIDIC and Chinese standards

Items	Conditions of China's construction contracts	Conditions of FIDIC construction contracts
Priority of documents	i. The appendix to contract ii. Standards, criteria iii. Design documents, files, and drawings	i. The client's requirements ii. The schedules iii. The contractor's proposal
Languages	Documents are written and interpreted in simplified Chinese. When more than one language is used in the particular conditions, *Chinese* is the ruling language.	If there is more than one language used, the version stated as the ruling language in the appendix to the tender shall prevail.
Base date	*30 days* prior to the latest date for submission of the tender	28 days prior to the latest date for submission of the tender.
Client's duty	Not stated	Client's financial arrangements are stated.
Engineer's authority	Stated in the particular conditions	The engineer may exercise the authority attributed to them as specified or necessarily implied from the contract.
Performance security	Not stated	The contractor shall obtain a performance security for proper performance, in the amount and currency stated in the appendix to the tender.
The contractor's representative	The contractor's representative shall be the contractor's personnel.	The contractor's representative can delegate any power, function, and authority to any competent person, and can at any time revoke the delegation.
The responsibility for data	*The client* shall be responsible for the accuracy and integrity of all data.	*The contractor* shall be responsible for the accuracy and integrity of all data.

Design and execution commencement date	The contractor shall start *on the date stated in the contract*.	The contractor shall start *as soon as is reasonably practicable after the date* stated in the contract (within 42 days after the contractor receives the Letter of Acceptance).
Right to vary	Upon receiving a variation notice, *the client* shall cancel, confirm, or vary the instruction.	Upon receiving a variation notice, *the engineer* shall cancel, confirm, or vary the instruction. The engineer may initiate variations at any time prior to issuing the Taking-Over Certificate for the works.
Staff and labor	Not stated	The rates of wages, working hours, facilities for staff, and labor health and safety are stated.
Obligations to the surroundings	Not clearly stated Mostly, the protection and maintenance of the construction site are regulated by the local administrative regulations.	The general obligations of the contractor shall include the avoidance of interference, the prevention of disorderly conducts, the protection of the environment and access route, the security of the site.
Subcontractor	Unless otherwise stated in the particular conditions, the proposed subcontractors shall obtain the prior consent of *the client*.	Unless otherwise stated in the particular conditions, the proposed subcontractors shall obtain the prior consent of *the engineer.*
Time program	The submission date and the number of copies for the time program prepared by the contractor are stated in the particular conditions.	The contractor shall submit a detailed time program within 28 days after receiving the notice under subclause 8.1 [Commencement of Works]. Unless otherwise stated in the particular conditions, monthly progress reports shall be prepared by the contractor and submitted to the engineer in six copies.

(*Continued*)

Table 5.2 (Continued)

Items	Conditions of China's construction contracts	Conditions of FIDIC construction contracts
Taking over of parts of the works	Not stated	The client shall not use any part of the works unless and until the engineer has issued a Taking-Over Certificate for this part. If a Taking-Over Certificate has been issued for a part of the work, the delay damages for completion of the remaining works thereafter shall be reduced.
Daywork	Not stated	For work of a minor or incidental nature, the engineer may instruct that a variation shall be executed on a daywork basis. The work shall then be valued in accordance with the daywork schedule included in the contract.
Tests after completion	A joint leading group formed by the client is in charge of the tests and all preparation works under the guidance of the contractor's personnel.	If the contractor does not appear on the time and place agreed, the client can proceed with the tests after completion, which will be deemed to be made in the contractor's presence, and the contractor shall accept the readings as accurate.
Valuation method	The two contracting parties may use *three kinds of contracts with different valuation methods* (fixed price contract, adjustable price contract, and cost reimbursement contract).	*Unit price contract* is used. The contract price shall be the Lump Sum Accepted Contract Amount, which is not changed during the implementation of the contract. The client shall pay in accordance with the measured amount.
Terms of service	Foundation and structural engineering (regulated in the design documents)	Auxiliary parts (half a year)
	Prevention of leakage (5 years) Equipment and fitting-out works (2 years) The client and the contractor could determine the others.	The main engineering and equipment (1 year). Important devices (a year and a half). The defects notification period shall not be extended by more than 2 years.

Advance Payment	The specific amount would be stated in the particular conditions and shall be paid to the contractor in a lump sum. The method, ratio and timing of the deductions of advance payment shall be stated in the particular conditions.	The total advance payment, the number and timing of installments (if more than once) shall be stated in the appendix to the tender. The advance payment shall be repaid through percentage deductions in the payment certificates.
Issue of payment certificates	Not stated	The engineer shall, within 28 days after receiving a statement and supporting documents, issue to the client an Interim Payment Certificate. Accordingly, the Final Payment Certificate will be issued later.
Disputes	If a dispute arises between the parties, both parties should at first try to settle the dispute amicably. Unless mediated by the local authority, the dispute shall be finally settled by arbitration or by a lawsuit.	If a dispute arises between the parties, either party may refer the dispute in writing to the DAB for its decision. Unless settled amicably, any dispute in respect of which the DAB's decision (if any) has not become final and binding shall be finally settled by international arbitration.
Payment of retention money	*Within 15 days* after the expiry date of the defects notification period, the outstanding balance of the retention money shall be certified.	*Promptly* after the expiry date of the defects notification periods, the outstanding balance of the retention money shall be certified.

5.2.2 Procurement Methods

Procurement systems (also called contractual management) are organizational processes where specific responsibilities and authorities are assigned to different stakeholders. Procurement also defines the relationships between project elements (Love, Skitmore, & Earl, 2010). The client has different contracting strategies to choose from (Fig. 5.4): traditional (design, bid, build); design and construct; and management. Each procurement method has its own benefits and drawbacks, and thus comprehensive understanding is needed when choosing one over the other. Fig. 5.5 shows how risk is distributed between the client (employer) and contractor. Speculative risk corresponds to the risk that can be distributed between parties in a contract (CRC Construction Innovation (CI), 2008).

5.2.2.1 *Traditional Procurement*

The traditional procurement method (Table 5.3) follows the common procurement route of design, bid, and build. The design and construction works are distinct and thus carried out by two separate teams (each having to perform their own cost-control). The contractor work can also be further subdivided (Fig. 5.6). The latter is responsible for carrying out all

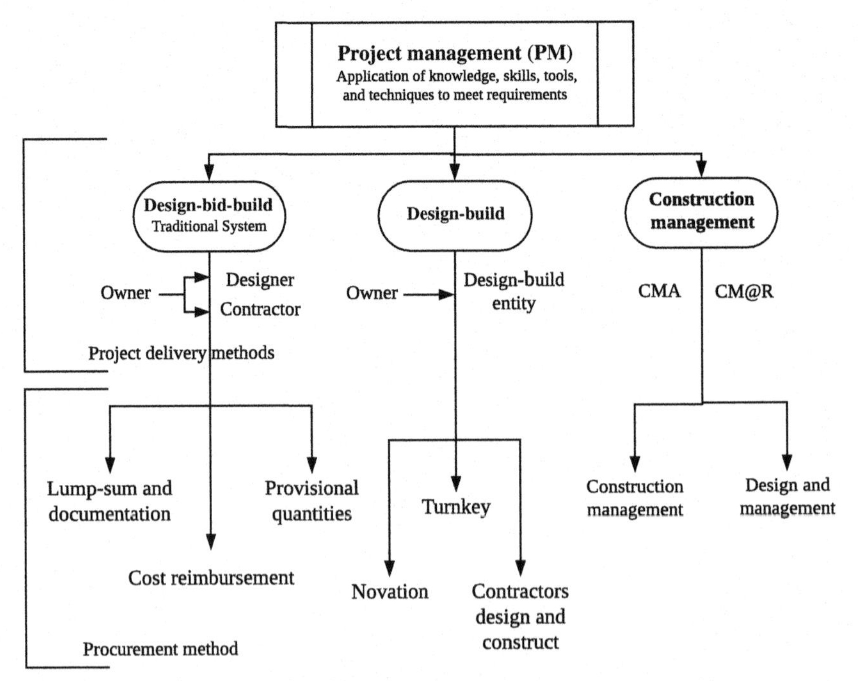

Figure 5.4 Categorization of building procurement systems.

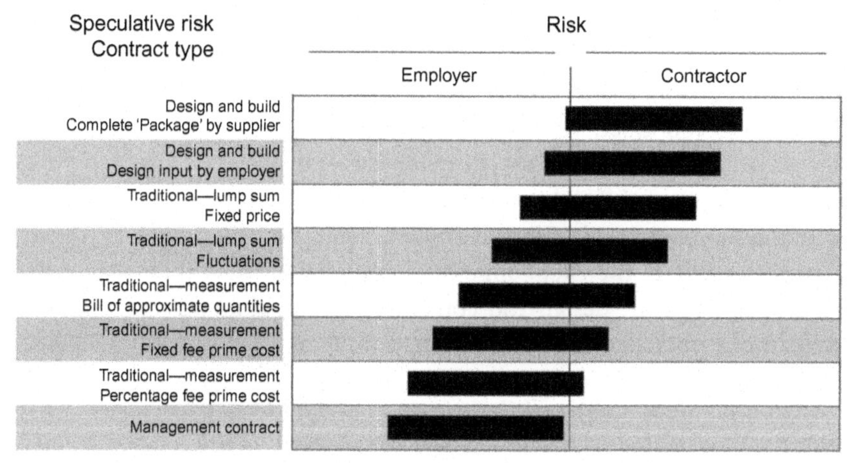

Figure 5.5 Risk appointment between client and contractor (CI, 2008).

Table 5.3 Pros and cons of traditional procurement

Pros	Cons
• Competitive equity as contractors bid on same basis • Design lead so client can have direct influence; improves overall design quality • Price certainty at the contract award • Contract changes easy to deal with • Market used to this procurement method	• The process can be time consuming as tender documents must be precise • Construction cannot start before design completion • Contractors cannot influence design

Source: Data from CRC Construction Innovation. (2008). *Report—Building procurement methods.* Retrieved from <http://www.construction-innovation.info/images/pdfs/Research_library/ResearchLibraryC/2006-034-C/reports/Report_-_Building_Procurement_Methods.pdf>.

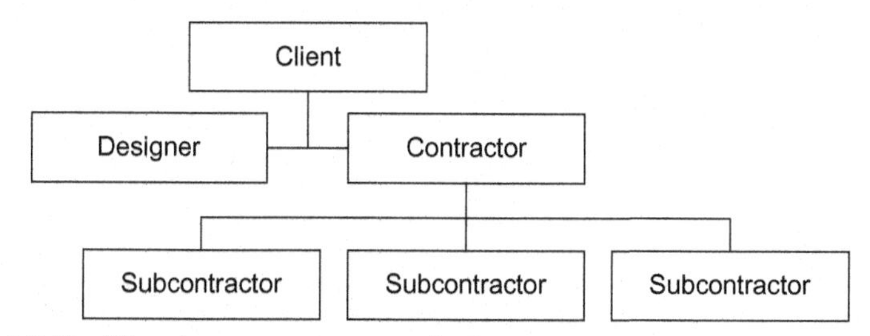

Figure 5.6 Traditional procurement organization.

tasks related to workmanship, materials, and those of the subcontractors (CI, 2008). Depending on the financial structure of the contract, traditional procurement can further be divided into a lump sum contract, measurement contract, and cost reimbursement.

5.2.2.2 Design and Build Procurement

In the design and build (also called design and construct, Table 5.4) procurement method, the client appoints a contractor that will be responsible for all or part of the design and construction tasks (Fig. 5.7). If the client wants more influence over the design, consultants can prepare a concept design and outline specifications before the contractor begins work.

Table 5.4 Pros and cons of design and build procurement

Pros	Cons
• The client has to deal with only one firm and thus needs less resources and time for contracting • Price certainty is obtained before construction • The potential use of a guaranteed maximum price with a savings option can enhance innovation and reduce time and cost • Overlap of design and building can shorten the project duration • Constructability is improved as the contractor can influence the design	• Client has to prepare an adequate and sufficiently comprehensive procurement which can be troublesome • Changes can be expensive • Bidding comparison can be difficult as the bidders' project can be completely different • The client has to commit to a concept design at early stages and before detailed designs are complete • Design liability is restricted to the available standard contracts

Source: Data from CRC Construction Innovation. (2008). *Report—Building procurement methods*. Retrieved from <http://www.construction-innovation.info/images/pdfs/Research_library/ResearchLibraryC/2006-034-C/reports/Report_-_Building_Procurement_Methods.pdf>.

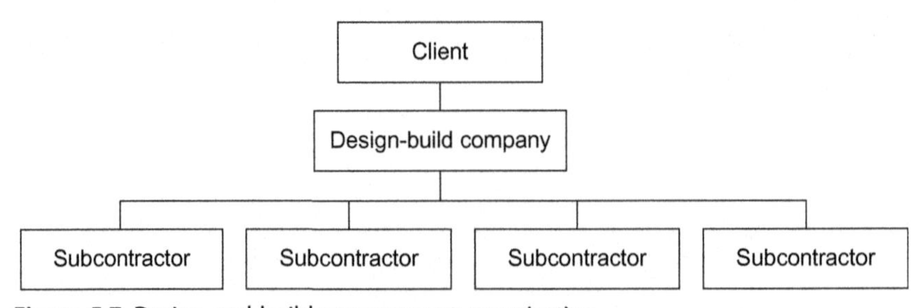

Figure 5.7 Design and build procurement organization.

Design and build is an integrated approach that passes design risks and responsibilities onto the contractor. The drawback for the client is that it has less influence on design and quality.

5.2.2.3 Management Procurement

Management procurement (Table 5.5) has different variants. The two most common are management contracting and construction management. They involve appointment of a project manager (or management contractor) by the client (Fig. 5.8). The early involvement of a project manager is favorable, as it improves the design constructability and management effectiveness. There are two differences between management contracting and construction management. For the latter (Gehr, 2013):

Table 5.5 Pros and cons of management procurement

Pros	Cons
• The client has to deal with only one firm and thus there is good coordination and collaboration between designers and constructors. • Overlap of design and building can shorten the project duration. • Work packages can be set competitively at prices that are current. • Constructability is improved as the contractor can influence the design. • Flexibility for changes in design.	• Price uncertainty. • Client has to be informed and proactive. • Close time and information control required. • Client must give a good quality brief to design team as design will only be completed once resources are committed to the project.

Source: Data from CRC Construction Innovation. (2008). *Report—Building procurement methods*. Retrieved from <http://www.construction-innovation.info/images/pdfs/Research_library/ResearchLibraryC/2006-034-C/reports/Report_-_Building_Procurement_Methods.pdf>.

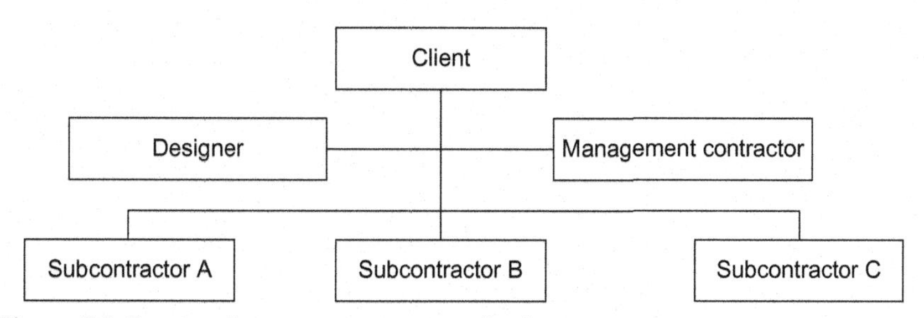

Figure 5.8 Construction management organization.

- The project manager only acts as a manager and thus does not have direct contracts with the contractors.
- The project manager is responsible for the design team selection.

5.2.3 Contract Documentation

Contract documentation consists of all documents needed to form a contract. These documents must provide enough information to allow contractors to complete their work. The principal documents included contract documentation are (Master Builders, n.d.):

- contract—the binding document between the client and the contractors;
- contract conditions—definition of the legal rights and obligations;
- special contract conditions;
- bills of quantities—list of materials, parts, and labor with their costs;
- drawings—architectural and structural plans;
- specifications—a project's technical requirements; and
- other documents (such as technical and pricing schedules).

Concerning specifications, they are either performance specifications (open) or prescriptive ones (closed). With the former, the contractor must carry out further design modifications. With the latter, the contractor cannot change the design.

5.2.4 Claim and Dispute Resolution

A dispute occurs when a disagreement between the client and the contract occurs. Claims result from disputes and are one party's formal notice to the other regarding the matter. The American Federal Acquisition Regulation defines a claim as a written demand/assertion by one of the contracting parties requesting (Federal Research Division, 2014):

- payment of a specific sum of money,
- adjustment or interpretation of contract terms, and
- other relief related to a contract.

Disputes and claims are almost unavoidable between two contractual parties, and thus it is important that proper dispute resolution processes are adopted. Disputes should be resolved in accordance with the contract and in a prompt and cost-effective manner.

5.2.5 Bonds

A bond is a legal means of protection against an adverse event that leads to noncompletion of a project. Backed by a surety company, there are different kinds of bonds (Table 5.6).

5.3 RISK MANAGEMENT

5.3.1 Static Risk Management Method

A risk is a situation involving exposure to a danger that could have consequences to safety or health, and impact on the environment, and/or effects on the project's costs. It is given by the product of the occurrence probability of a defined hazard and its occurrence

$$Risk = \sum_i probability(P) * consequence(C) \tag{5.1}$$

Tunneling projects are high-risk projects as geology and geotechnical ground properties are never fully known. A report from the Construction Industry Research and Information Association (1978, cited by Muir Wood, 2000) summarizes the principal causes of risk for tunneling projects (Fig. 5.9).

It is crucial to manage risk to avoid unwanted surprises. Risk management is a core task of working teams in tunneling projects. It minimizes undesired risk and increases opportunities for the most appropriate and innovative designs to succeed.

Table 5.6 Different types of bonds

Type of bond	Descriptions
Performance bond	A means of insuring a client against the risk of a contractor failing to fulfill contractual obligations.
Advance payment bond	A means for the client to secure an advance payment to the contractor.
Off-site materials bond	A bond used when the client agrees to pay for items even when they remain off-site.
Bid bond	A potential requirement for international tender processes to secure the tender's commitment to start the contract.
Retention bond	A means used when the client agrees to pay the retention before the proper completion of a contractor's activity.

Source: Data from Designing Buildings (2017). *Bonds in construction contracts*. Retrieved from <https://www.designingbuildings.co.uk/wiki/Bonds_in_construction_contracts>.

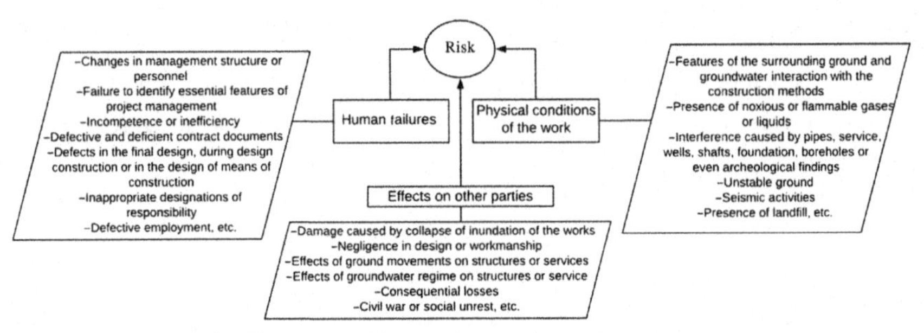

Figure 5.9 Principal risks in tunneling projects.

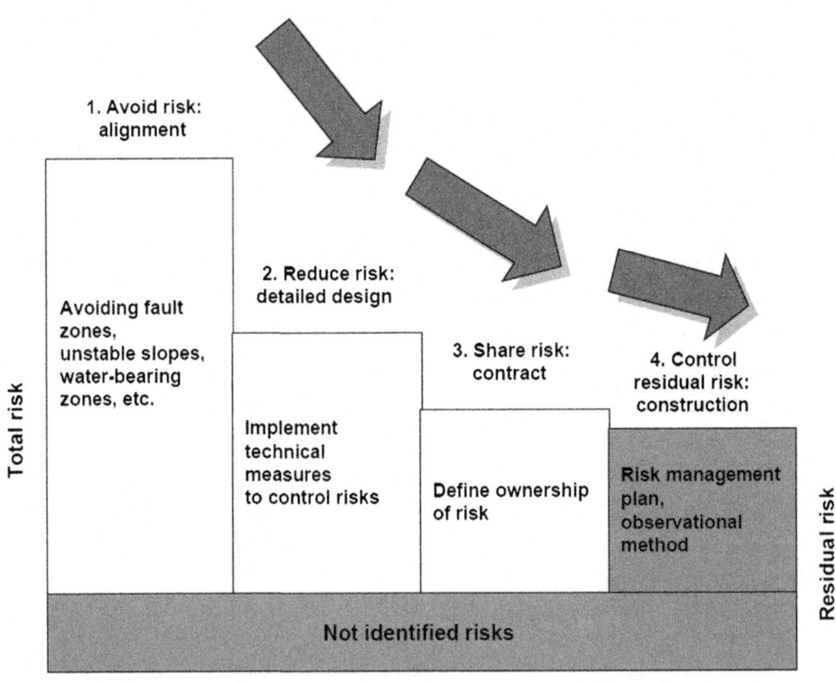

Figure 5.10 Risk management and risk reduction during a project (Schubert, 2006).

A thorough risk analysis is needed for all projects, as each project is unique and presents its own risks. It is therefore impossible to give an exhaustive list of all the causes of underground construction risks. All parties involved in a project must have clearly defined responsibilities with regard to risk.

Fig. 5.10 breaks down the process of risk management into four major phases (Schubert, 2006): risk avoidance, risk reduction, risk sharing, and residual risk control. Each of these phases follows a cyclic process (Fig. 5.11).

Figure 5.11 Risk management process.

Thorough risk management through these four phases reduces risk to the minimum residual risk. The latter is never eliminated, irrespective of the quality of risk management. Each phase of Fig. 5.10 is detailed as follows.

1. Project phase 1 (the alignment design and preliminary part of a project such as geological prospecting) is when potential general risks are identified. The tunnel alignment is then optimized (Schubert, 2006). A risk acceptance criterion should also be set during this phase. Its purpose is to establish the minimum standards for risk assessment.
2. Project phase 2 (the detailed design) is when more detailed risk assessment takes place. This is possible because more information is available at this stage of a project and so the risks can be quantified more easily. The design is also then optimized such that the risks are minimized. Criteria for the observational method are also defined at this stage of a project (Schubert, 2006).
3. Project phase 3 (the tender process) is when identified risks are shared among the parties involved in the contract. This is clarified in the tender documents and during contract negotiation and ensures efficient risk management during construction.
4. Project phase 4 (the construction phase) is when the client and contractors develop their own risk management system to reduce the respective risks they are responsible for.

Tables 5.7 and 5.8 show how risks are evaluated and how monitoring methods are adopted. Risks are categorized into five levels using these tables, according to risk occurrence probability and risk consequence. High risks (high occurrence probability and severe consequence) require stricter control. The economic losses arising from risk failure are listed in Table 5.9.

Table 5.7 A matrix of risk evaluation

Risk occurrence probability (from A to E, A being the lowest)	Risk consequence				
	1 Negligible	2 To be considered	3 Serious	4 Very serious	5 Catastrophic
A	G	G	B	Y	O
B	G	B	Y	Y	O
C	G	B	Y	O	R
D	B	Y	O	O	R
E	B	Y	O	R	R

Note: Color code explained in Table 5.8.

Table 5.8 Risk control policy

Level	Risk color (according to Table 5.6)	Control policy	Control schemes
1	Green (G)	Negligible	Daily management and survey
2	Blue (B)	Allowable	Attention is needed, daily management and survey should be strengthened
3	Yellow (Y)	Acceptable	Decision is needed, control and precaution measures are needed
4	Orange (O)	Unacceptable	Decision is needed, control and precaution measures must be made
5	Red (R)	Prohibited	Site shutdown state, corrective action

Table 5.9 Economic loss (EL, million RMB)

Risk consequence	Economic loss
1	EL < 5
2	5 ≤ EL < 10
3	10 ≤ EL < 50
4	50 ≤ EL < 100
5	EL ≥ 100

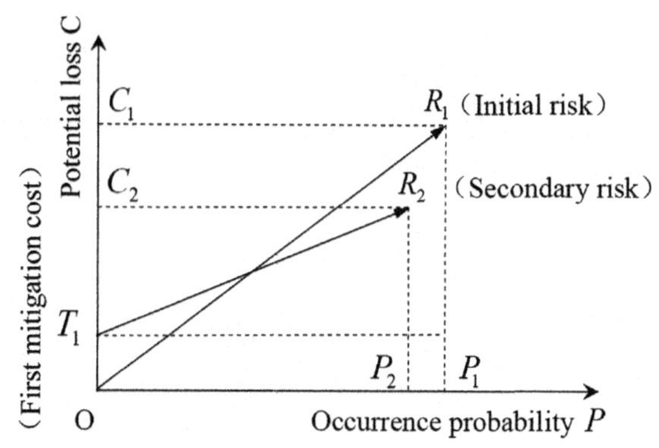

Figure 5.12 Initial risk and secondary risk.

5.3.2 Multiphase Risk Management Method

Most of the risk management methods such as those mentioned above were made using static analysis. However, during the risk management steps discussed, environmental factors may change, and deviations from the original design assumptions may also occur. New risks may be identified, and existing risks may be transferred or disappear. Risk items after mitigation should be reassessed for their likelihood of reoccurring and any possible impacts. Risk monitoring and control is a continuous process, that is, a dynamic risk evaluation process. But the effects or costs arising from risk reassessment and mitigation at different construction stages are usually not considered. This is due to the lack of dynamic formulae to quantify a project's risk in a dynamic and evolving environment. To achieve this purpose, a multiphase risk management method can be used.

In this method, the assessment of risks is divided into two phases: the initial risk phase and the secondary risk phase (Fig. 5.12).

The initial risk phase is used for risk evaluation at the beginning of a project and can be quantified by a combination of probability and loss. The initial risk is usually expressed as a combination of the risk probability (P_1) and the potential loss (C_1), and can be formulated as $R_1 = f(P_1, C_1)$. The initial risk is always put forward at the beginning of a project based on construction plans. It is completely foreseeable and can be well controlled with appropriate mitigation measures.

Secondary risk can be expressed as a combination of the risk probability (P_2), the potential loss (C_2), and the mitigation cost (T_1), in the form of $R_2 = f(P_2, C_2, T_1) = T_1 + f_2(P_2, C_2)$. From this formula, we know that the secondary risk is not only dependent on the risk after mitigation measures, but also on the cost of those mitigation measures. It follows that different mitigation measures have different mitigation costs (T_1). Risk after such mitigation measures are taken into account (f_2) and a secondary risk (R_2) will result. The secondary risk phase involves identifying risks after implementing the selected mitigation measures for the initial risk; the mitigation costs are also included. Obviously, the secondary risk must be lower than the initial risk. The task of tunnel engineers is to find out the mitigation measure with the lowest secondary risk. In other words, the target is to optimize the mitigation measure of the initial risk.

Based on the results of the risk evaluation, risks can be divided into three types: acceptable risk, acceptable risk after mitigation, and unacceptable risk.

If the initial risk is located at the lower-left corner in the green area of Fig. 5.13, it is acceptable. This kind of risk is allowed without any further mitigation. However, monitoring during construction is still necessary even in those projects deemed to have an acceptable risk profile, and appropriate emergency plans should still be prepared.

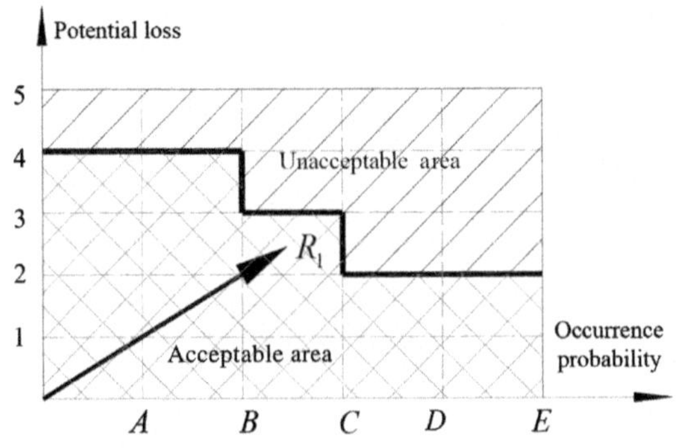

Figure 5.13 Acceptable risk.

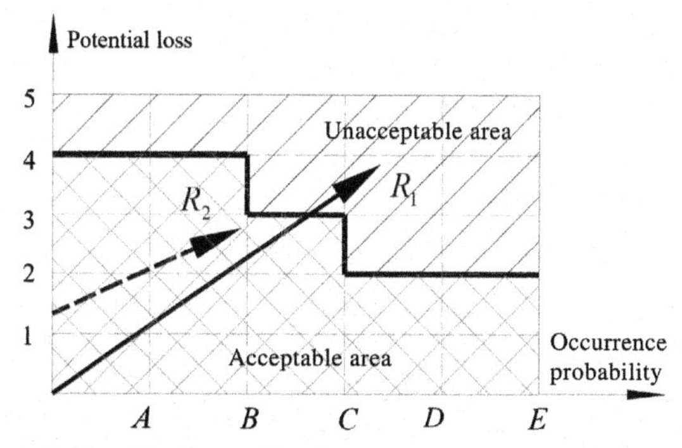

Figure 5.14 Acceptable risk after mitigation.

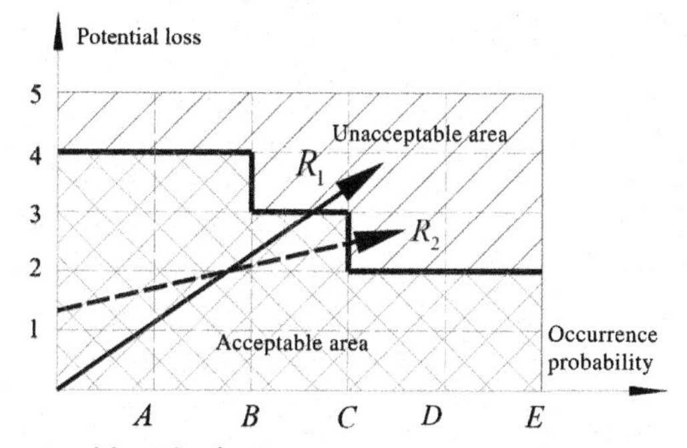

Figure 5.15 Unacceptable risk after mitigation.

If the initial risk is located at the upper-right corner in the red area in Fig. 5.13, it is unacceptable, and appropriate mitigation measures must be implemented. If the secondary risk after mitigation falls into the green area, as shown in Fig. 5.14, it is an acceptable risk after suitable mitigation.

If, however, the secondary risk after mitigation is still in the red area, as shown in Fig. 5.15, then it belongs in the unacceptable risk category, and the project cannot move forward.

This multiphase risk management approach can be applied to tunnel engineering as well as other engineering fields.

5.4 QUALITY, HEALTH, SAFETY, AND ENVIRONMENT PERFORMANCE

Clients have legal responsibilities that hold them accountable for the quality, health, safety, and environment (QHSE) of their project. The QHSE management of a project is governed by the country's primary legislation, case law, and tort. Identifying, assessing, and controlling risk, risk management can therefore be deemed a lead up to QHSE management.

5.4.1 Quality

International competition has made quality control a primary concern in industry. In addition to maintaining a company's reputation, problems concerning construction quality can also lead to lawsuits. A client must therefore properly define its quality requirements and ensure they are delivered.

Cost estimates and design and feasibility studies aim to find the optimal (economical and technical) design. To avoid costly changes during construction, it is crucial that quality control is carried out before any substantial work.

5.4.2 Safety

The construction industry (among which tunneling is not an exception) is high risk. In the United States, for instance, in 2015, a fifth of the worker deaths in industry happened in the construction industry (Occupational Safety and Health Administration, n.d.). It is therefore essential that safety be ensured on construction sites. All parties in a tunnel project must help ensure this.

Safety standards and procedures must be defined, and be carried out taking into consideration past experience and newer technologies that may be safer. Clean and tidy work sites can also prevent accidents, and areas with special needs or higher risk should be properly marked. Feedback from workers should also be considered as they can see dangers that may not have been known.

5.4.3 Environment

Tunnelers should work on making their projects environmental friendly for construction and operations phases. Considering the environmental impact while choosing a construction method is important since it can help decrease carbon emissions, etc. Taking material transportation and waste recycling as examples, certain more environmentally friendly operations come at a cost.

Tunnelers must consider the environment in the early stages of a project design and its organization. For example, warm air and water in drain tunnels can be used as energy reserves. This heat could be used for pace heating, greenhouses, balneology, wellness, fish farming, etc. (Rybach, 2010).

Green constructions are those projects that save on resources to the utmost extent and coexist with nature. To carry out such projects, a lot of technical expertise is needed to save resources and reach environmental protection goals.

5.4.4 Health

During construction, tunneling projects produce a lot of dust underground. For health reasons, ventilation and dust prevention mechanisms are put in place (e.g., humidity drilling, mechanical ventilation, individual prevention). Oxygen concentration, dust density, presence of poisonous gases, and noise exposure are also all monitored continuously.

Upon completion, a tunneling structure must also ensure end users have access to high-quality air (such as in metro lines) and water. Underground structures often used by the public should also ensure temperature regulation and proper illumination.

5.5 CASE STUDY: SHANGHAI METRO LINE NO. 4 RECOVERY PROJECT

5.5.1 Project Background

Twin bored tunnels were constructed between Pudong South Road Station and Pudong South Bridge Station for the Shanghai Metro Line 4. The main tunnels, including a 440-m section under the Huangpu River, were completed successfully. The construction of a cross-passage, which was also connected to the ventilation shaft right above, using ground freezing and mining excavation, started on July 1, 2003. During excavation, the cross-passage collapsed because of inadequate ground freezing. The surrounding soil, together with underground water, rushed into the working face and subsequently into the completed main tunnels. The severe ground loss and collapse of more than 200 m of the main tunnels resulted in the sinking of the ventilation shaft and the settlement of nearby buildings (Fig. 5.16). Cavity filling was carried out after the collapse as part of the damage-control measures. After the incident, a technical committee was set up to evaluate and determine possible remedial measures. The repair options fell into two categories, in situ remediation and tunnel realignment, as shown in Fig. 5.17. The in situ remediation was adopted after comprehensive consideration and analysis.

Figure 5.16 Tilting of a building above the collapsed area.

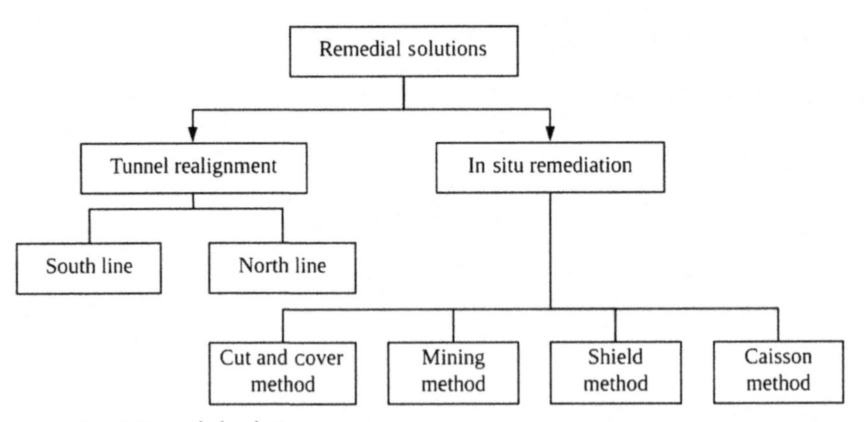

Figure 5.17 Remedial solutions.

5.5.2 Risk Analysis of Remedial Solutions

Four possible solutions were identified in the in situ remediation scenario: cut and cover; the New Austrian Tunneling Method (NATM), which is associated with ground reinforcement; shield excavation; and a continuous pneumatic caisson method. The risk analysis of each solution is shown in the following sections.

- *Cut-and-cover method*

  The damaged tunnel can be divided into three parts: the east tunnel, the central tunnel, and the west tunnel (Fig. 5.18). To facilitate the reconstruction of the damaged tunnels, parallel retaining walls 23 m away from the tunnels and some 263 m in length were constructed along the damaged tunnels. The excavation was 38 m in depth. At some sections, the excavation depth was up to 41.2 m. Apart from one section of the retaining wall in the west part that was excavated under the traffic decking to maintain the traffic flow, the rest was designed to be open-cut excavation from the surface. For the east part, where 60 m of the damaged tunnel was under the Huangpu River, a steel platform was erected in combination with a sheet-piled cofferdam to facilitate land construction. The risk analysis of the excavation works is shown in Table 5.10. Using data from Tables 5.11–5.13, the construction cost, initial risks, and relevant mitigation costs of the cut-and-cover solution were assessed, and these are shown in Table 5.14. From Table 5.15 it can be seen that the sum of the construction cost and risk without mitigation is much higher than the sum of the construction cost, mitigation cost, and risk after mitigation. Therefore, if the cut-and-cover method is adopted as the final remedial solution, the initial risk mitigation should be implemented.

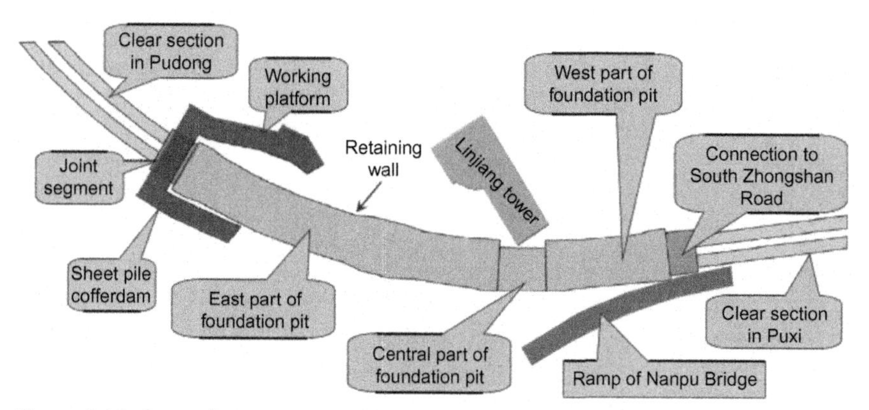

Figure 5.18 Cut-and-cover construction.

Table 5.10 Risk analysis of cut-and-cover construction

Initial risks and evaluation		Proposed technical measures	Secondary risks and evaluation	
Tunnel subsidence due to installation of cofferdam	3C	Cavity filling conducted as part of the damage-control measures.	Great difficulties in guaranteeing the uniformity and strengthening of deep foundation reinforcement.	3C
Cofferdam deformation	2D	Bored piles and lightweight cofferdam adopted to protect the intact tunnel.		
Protection of the surrounding environment	4B	Vertical ground freezing plug is formed at the interface between the damaged and intact tunnel prior to cofferdam construction.	Piping, quicksand, and the decline of the groundwater table may occur.	3B
Construction of deep foundation pit	3D	Oscillatory drilling and cutting machine with high accuracy is adopted for the removal of damaged segments.	High risks during construction of the retaining wall, excavation of foundation pit, and dewatering.	
Reinforcement of deep foundation	3B			
Dewatering of deep confined water	3C	1.2 m thick diaphragm wall with "T" section steel is adopted to enhance water tightness.	High risks in security of the tightness of the vertical ground freezing plug at the interface between the damaged tunnel and intact tunnel.	3B
		Reinforced concrete strutting system is adopted to increase the rigidity of the retaining wall. Sacrificial jet grout layers below the strut level and the foundation level is installed. A series of recharging wells are placed strategically to safeguard the surrounding structure against dewatering effect.		

Table 5.11 Construction cost, risks, and mitigation cost of cut-and-cover construction

Without mitigation Construction cost	Risk	Cost of initial risk mitigation	With mitigation Construction cost	Risk
30	0.165	5.11	30	0.165
7.5	0.413			
75	0.041	28.95		
30	1.650	19.64	30	0.0165
30	0.017	5.66		
30	0.165	2.94	30	0.0165
75	0.041	17.43		
30	0.017	2.28		
307.5 (total)	2.51 (total)	82.01 (total)	90 (total)	0.198 (total)

Table 5.12 Risk analysis of continuous pneumatic caisson construction

Initial risks and evaluation		Proposed technical measures	Secondary risks and evaluation	
Construction of cofferdam and soil backfilling	3C	Soil mixing wall method and oscillatory jet grouting piles with different lengths are adopted to protect the environment.	Ground settlement may be induced by caisson sinking.	4C
Protection of the surrounding environment	4C			
Connection of adjacent caissons	4D			
Construction of caissons under great buried depth	3C	Suitable sealing wall for connecting the caissons is adopted to minimize ground disturbance. Application of compressed air to prevent the tilting of the caissons and ground loosening.	Ground loosening and settlement may be induced between the connecting walls and adjacent caissons.	4D
Connection of the caissons to the intact tunnel	4D			
Lack of experience	4D			

Table 5.13 Construction cost and risk of continuous pneumatic caisson construction without mitigation

Initial risk	Construction cost	Risk
(1)	30	0.165
(2)	75	0.4125
(3)	75	4.125
(4)	30	0.165
(5)	75	4.125
(6)	75	4.125
Total	360	13.12

- *Continuous pneumatic Caisson method*

  Five main construction steps are involved in the continuous pneumatic caisson method (Fig. 5.19): 1. construction of the cofferdam in the Huangpu River and soil backfilling; 2. sinking of the first three caissons (Nos 1, 3, and 5); 3. sinking of the remaining two caissons (Nos 2 and 4); 4. connection of the adjacent caissons; and 5. connection of the caissons to the intact tunnel. The risk analysis of the caisson construction is shown in Table 5.12. From Tables 5.10 and 5.11, the construction cost and risk of the continuous pneumatic caisson method were evaluated, and are shown in Table 5.13.
- *NATM with ground reinforcement*

  The NATM with compressed air support and ground improvement is adopted frequently in the construction of cross-passages in soft soil. The relevant technologies are well-developed and proven. The main construction process is shown in Fig. 5.20. The main risks and corresponding evaluations are shown in Table 5.14. From Tables 5.10–5.12, the construction cost and risk of the NATM were assessed, as shown in Table 5.14.
- *Shield method*

  The shield method has been adopted widely in subway construction in soft soil. However, for remedial construction, the surrounding environment and the geotechnical conditions had already been changed. Some tunneling processes would therefore have to be adjusted. The main construction process is shown in Fig. 5.21, and the main risks and their corresponding evaluations are shown in Table 5.16. From Tables 5.10, 5.11, and 5.16, the economic loss of the shield method can be evaluated, as shown in Table 5.17.

Table 5.14 Risk analysis of NATM construction

Initial risks and evaluation		Proposed technical measures	Secondary risks and evaluation
1. Protection of surrounding environment	4C	1. Oscillatory drilling and cutting machines with high accuracy are adopted for the removal of damaged segments.	1. Differential ground settlement may occur during dewatering.
2. Tunnel clearing	3B		
3. Quicksand during excavation	4D	2. Bench method and full section grouting in advance is adopted during mining.	2. Pipeline, quicksand, and failure of the excavation face may be induced by poor ground reinforcement.
4. The reinforcement of deep foundations	3B	3. Triple tube jet grouting technology is adopted for foundation reinforcement.	
5. The dewatering of deep confined water	3C	4. Well-point dewatering and a series of recharging wells are placed strategically. 5. Vertical ground freezing plug is formed to isolate the damaged tunnel from the intact tunnel.	
6. Stability of the tunnel excavation face	5B		3. Large surface settlement and excessive ground loss into the excavating face may be induced by excavation.

Table 5.15 Construction cost and risk of NATM construction without mitigation

Initial risk	Construction cost	Risk
(1)	75	0.4125
(2)	30	0.0165
(3)	75	4.125
(4)	30	0.0165
(5)	30	0.165
(6)	100	0.055
(7)	75	0.4125
Total	415	5.203

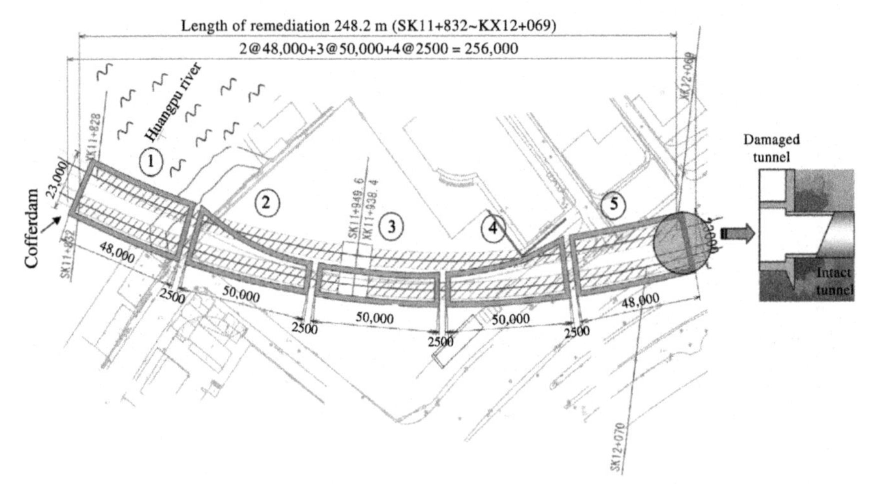

Figure 5.19 Continuous pneumatic caisson construction.

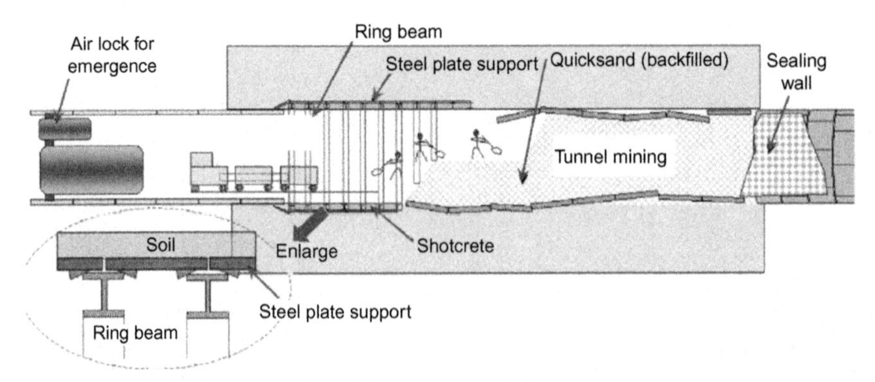

Figure 5.20 NATM construction with compressed air support.

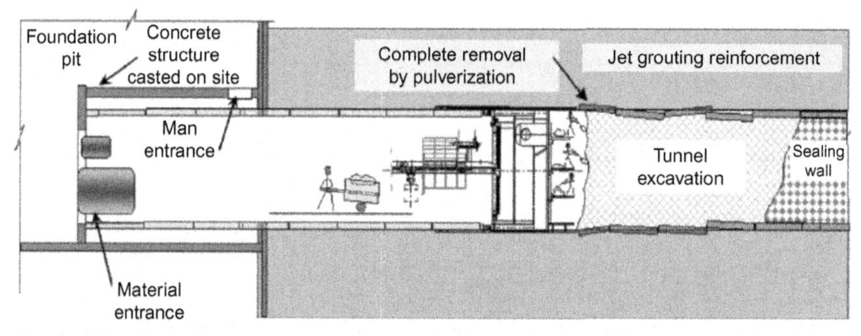

Figure 5.21 Shield construction with compressed air support.

5.6 QUESTIONS

5.1. For a specific underground project, how does the design stage differ from the construction stage? In which stage would you prefer to be involved, and why?

5.2. Why is the QHSE management of a project important?

5.3. Give your opinion about the difference between Chinese Standards and FIDIC shown in this chapter.

5.4. Explain the differences and similarities of FIDIC and NEC.

5.5. Define the terms "life cycle" and "life cycle assessment." Why is it important to apply these two concepts and how can they help create sustainable development in underground engineering?

5.6. Describe three types of traditional procurement methods and their advantages and disadvantages.

5.7. Why do we need risk management in underground engineering? What is the principle?

5.8. Complete the following table.

Tunneling project management	Duties and responsibilities
Owner's Tunneling Project Management	
Designer's Tunneling Project Management	
Contractor's Tunneling Project Management	

5.9. Why do we need risk management in underground engineering, and what are the key issues to be considered compared to engineering projects at the surface?

5.10. What is the principle of risk management?

Table 5.16 Risk analysis of shield method

Initial risks and its evaluation		Proposed technical measures	Secondary risks and its evaluation	
1. Protection of surrounding environment	3B	1. Oscillatory drilling and cutting machines with high accuracy are adopted for the removal of damaged segments.	1. High risks exist in shield launch and reception in sandy formation at deep depth.	5C
2. Tunnel clearing	3B	2. Vertical ground freezing plug is formed to isolate the damaged tunnel from the intact tunnel.	2. Excessive ground loss, difficulties in shield control, and hand ring build may be encountered during tunneling in sandy formation with confined water and through the changed conditions of the damaged tunnel section.	5B
3. Construction of water pumping station and cross-passage	4C	3. The ground freezing method is adopted for shield launch and reception.		
4. Shield launch and reception at depth	4D			
5. Shield tunneling in sandy formation	4B			
6. Connection of the new tunnel to the sound tunnel	3C			

Table 5.17 Construction cost and risk of shield construction without mitigation

Initial risk	Construction cost	Risk
(1)	30	0.0165
(2)	30	0.0165
(3)	75	0.4125
(4)	75	0.4125
(5)	75	0.04125
(6)	30	0.165
Total	315	4.78

REFERENCES

CRC Construction Innovation. (2008). *Report—building procurement methods*. Retrieved from <http://www.construction-innovation.info/images/pdfs/Research_library/ResearchLibraryC/2006-034-C/reports/Report_-_Building_Procurement_Methods.pdf>.

Designing Buildings. (2017). *Bonds in construction contracts*. Retrieved from <https://www.designingbuildings.co.uk/wiki/Bonds_in_construction_contracts>.

Dickson, R. N. (2013). *An analysis of the use and implementation of NEC versus traditional forms of contract in the HK construction industry*. University of Bath, Department of Architecture and Civil Engineering.

Federal Research Division. (2014). *Chapter 22—Contract disputes act (CDA) claims*. Retrieved from <https://www.loc.gov/rr/frd/Military_Law/pdf/CAD_2014_Ch22.pdf>.

Gehr, F. (2013). *The difference between management contracting and construction management*. Retrieved from <http://tipsdiscover.com/home/difference-management-contracting-construction-management/>.

Love, P. E. D., Skitmore, M., & Earl, G. (2010). Selecting a suitable procurement method for a building project. *Construction Management and Economics*, *16*(2), 221–233.

Master Builders. (n.d.). *Contract documentation*. Retrieved from <https://www.mbqld.com.au/contracts-and-disputes/contract-documentation>.

Muir Wood, A. (2000). *Tunneling: Management by design*. London, England: E & FN SPON.

Musaka, T.J. (2016). *The case for standard forms of construction contract* [PDF]. Retrieved from <https://www.slideshare.net/TomJosephMukasa/isucaseforstandardformsofconstructioncontractfinal>.

NEC. (n.d.). *About NEC*. Retrieved from <https://www.neccontract.com/About-NEC>.

Occupational Safety and Health Administration. (n.d.). *Commonly used statistics*. Retrieved from <https://www.osha.gov/oshstats/commonstats.html>.

Pinsent Masons. (2011). *Standard form contracts: FIDIC*. Retrieved from <https://www.out-law.com/en/topics/projects--construction/construction-standard-form-contracts/standard-form-contracts-fidic/>.

Ritz, G. J. (1994). *Total construction project management* (1st ed). New York: McGraw-Hill.

Rybach, L. (2010). Geothermal use of warm tunnel waters—Principles and examples from Switzerland. *Transactions Geothermal Resources Council*, *34*, 949–952.

Schubert, P. (2006). Geotechnical risk management in tunneling. In A. C. Matos, L. R. Sousa, J. Kleberger, & P. L. Pinto (Eds.), *Geotechnical risk in rock tunnels* (pp. 53–62). London, England: Taylor & Francis Group.

CHAPTER 6

Operation Systems in Underground Engineering

Contents

After construction completion, an underground infrastructure enters its operation stage at which several issues must be addressed. These include ventilation, power and water supply among others. It is indeed important to ensure a certain comfort for users as well as the good and safe operation of the structure. Therefore, the design of a tunnel includes designing these operation systems and ensuring their overall performance when a tunnel is in use. To ensure safety, it is also crucial to manage risks in order to avoid or at least mitigate disasters and possible loss of life and property. This chapter first addresses the main principles for the most important operation systems before discussing certain disasters that risk occurring in underground structures. Finally, fire risk design is studied in more in depth through a case study.

Underground Engineering
DOI: https://doi.org/10.1016/B978-0-12-812702-5.00006-2

6.1 OPERATIONAL VENTILATION METHODS

This section focuses on operational ventilation techniques since proper ventilation is needed to ensure a safe and healthy working environment during construction as was briefly described in Chapter 4, Underground Construction, and Chapter 5, Project Management.

Ventilation methods for in-service underground structures vary with structure usage. For example, in a subway tunnel, ventilation can be accomplished by the mere movement of a train, as when it moves at high speed air is pushed and pulled out of the tunnel like a plunger. If a tunnel is too long to create this "plunger effect," middle ventilating shafts should be added.

Ventilation systems have large dimensions and can be complex systems depending on the different methods used. There are three types of operational ventilation (Fig. 6.1):

1. Longitudinal ventilation: the air path is from one cave to another with polluted air and dust carried in the longitudinal direction.

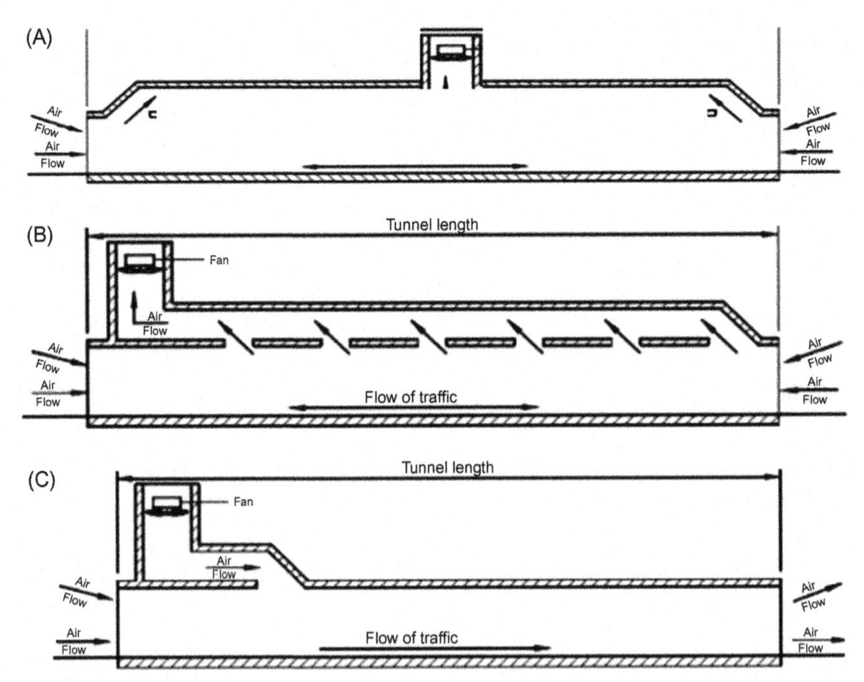

Figure 6.1 The three types of operational ventilation: (A) longitudinal ventilation by central extraction; (B) smoke extraction and intermediate ceiling ventilation; and (C) ventilation with jet fans.

2. Transverse ventilation: fresh air is pumped into the tunnel by air blowers and drawn out by holes on the sidewalls of the tunnel.
3. Half-transverse ventilation: fresh air is pumped into the tunnel through a single air tunnel and the waste air and dust are squeezed out of the tunnel in the longitudinal direction.

Longitudinal ventilation is widely used and relatively inexpensive as well as energy saving. One of its main drawbacks is that it can lead to waste accumulating at the extremities of the tunnel. From a safety point of view it is also troublesome in the case of a fire emergency, as fire can easily travel through longitudinal ventilation. This makes rescue efforts and fire extinguishing more difficult for firefighters.

Compared with longitudinal ventilation, transverse ventilation is more expensive due to its high efficiency at handling fire emergencies and waste. Indeed, a fire can quickly be extinguished and the waste air uniformly diluted rather than accumulated.

Lastly, half-transverse ventilation is less expensive than transverse ventilation due to its comparatively lower efficiency and functionality. This kind of ventilation is suitable for tunnels with a length of 1–3 km.

Tunnels should be equipped with the necessary technology to activate/deactivate ventilation, especially longitudinal ventilation, in order to prevent fire and especially smoke from traveling along the tunnel in the same direction as people leaving the tunnel during a fire event. Furthermore, an actuation protocol should be established before the tunnel is put to use so workers know how to use ventilation equipment in emergencies.

6.2 POWER SUPPLY

As with ventilation, power supply methods depend on the underground structure's function and dimensions, since different amounts of power are needed for lighting, ventilation, drainage power, environmental control, escalators, etc.

There are two types of power configurations: centralized and distributed. According to past experience and specifications, distributed power supply is used for long powerlines and concentrated power for short ones.

6.3 LIGHTING SYSTEM AND BRIGHTNESS

Underground lighting is either normal lighting or emergency lighting. The former is usually adopted in device rooms, control rooms, and public places. This kind of lighting uses standard devices as long as the normal demands can be met. Emergency lighting is a special lighting and includes reserve lighting, evacuation lighting, and safety lighting. This kind of lighting is always controlled by the emergency prevention system and must always be in good condition in case of an emergency.

The brightness inside a tunnel must carefully be controlled. For a road tunnel, the contrast inside and outside a tunnel can be difficult for drivers to handle. It can be dangerous, as their eyes have to get used to the contrasting luminosity at the entrance and exit of a tunnel. Dust and waste air in the tunnel can also cause poor visibility. It is therefore of paramount importance to properly design the lighting system for the safe operation of traffic in the tunnel. This applies during the day and at night. It is often achieved by installing appropriate lighting devices at certain positions to slow down the changing rate of brightness (usually 10:1 or 15:1 is acceptable). According to the Chinese Design Code for Underground Lightening [CECS45-92], brightness should be designed linearly in the tunnel, and the level of uniformed brightness should be less than one third of the average brightness. The light at the exit should be about five times the basic lighting level in the tunnel to reduce the "brightness difference effect."

6.4 WATER SUPPLY

Depending on the function of an underground structure, different volumes of water will have to be handled. In underground space water is distributed in different directions. Groundwater and used water should be collected separately and dealt with depending quality. For instance, used water is usually pumped into the nearest municipal sewer and rainwater is pumped into rainwater pipes.

The main principles for dealing with water supply include:

- Water should be saved and fully used.
- Sufficient amounts of water should be kept for daily use.
- City tap water is the main source of water.
- Water equipment must be kept in good condition and maintain high efficiency.
- Water quality must meet the international or national standards for drinking and operation.

6.5 SURVEILLANCE AND CONTROL SYSTEMS

6.5.1 Environmental Health Standards for Tunnel Operation

For underground structures, three kinds of parameters should be monitored:

- temperature and moisture level;
- gas proportion in the air (CO_2, O_2, NO, and NO_2); and
- particle matters with diameters of 2.5 μm and smaller ($PM_{2.5}$).

The temperature range that should be maintained depends on the location and time of the year as climate and people's varying preferences result in different definitions of comfort. The Chinese practice usually defines the suitable temperature of the structure to be from 16°C to 27°C while the relative moisture level that suits most people ranges from 40% to 60%.

Concerning the air-quality level, CO_2, CO, and O_2 are always used as indicators to verify that ventilation systems are supplying enough fresh air inside the structure. Research shows that the amount of CO_2 in volume should be maintained at 0.07%–0.15%, and CO should be no more than $1/10^4$ (Li, Li, & Li, 2010). Additionally, there should be some negative-oxygen ions inside the underground structure as they are necessary for human metabolism.

6.5.2 Control Systems

Besides the environmental health standards that ensure a comfortable and safe environment for people, several operation systems are necessary for the sound operation of a structure. Indeed, there are systems to monitor and control lighting, directing signals, alarms as well as surveillance devices to name but a few.

Also, mainly for energy-saving purposes, an important feature for some subways is to have a sealing system to isolate stations from the outside. Indeed, as passengers in subway tunnels are inside subway trains it can be useful to separate subway tunnel air from the station air (Zhang & Zhu, 2005):

- Open system: fresh air is interchanged in and out of the tunnel by the "valve effect" of the train and mechanical ventilation. Station air communicates with tunnel air. This system mainly applies to small passenger train networks and where the maximum temperature is less than 25°C.

- Closed system: the inner space is closed off from the exterior space with the use of platform screen doors. Some use ventilation systems for the tunnels while others don't. The advantage of the former is that the temperature difference between stations and running tunnels is not as big. Therefore, less air conditioning is required for the stations, reducing the overall noise level and making it more comfortable for passengers.

6.6 SAFETY AND DISASTER CONTROL

Underground facilities may need more consideration than surface facilities with regards to fires and other possible disasters during their design, construction, and operation. Engineers should therefore properly manage risks in order to avoid damage and loss of lives. Indeed, on the one hand, surrounding ground can help underground spaces resist attacks from the outside. On the other hand, underground spaces are confined and sometimes enclosed, which makes evacuation difficult with limited exits. Consequently, reliable safety control for underground spaces is needed.

Disasters are primarily natural or manmade. The former includes flooding, earthquakes, and slope failure. The latter includes fire, traffic accidents, terrorist attacks, and engineering failures. In reality, one catastrophe may consist of both types of disasters. For example, an earthquake may cause fires, explosions, and slope failure.

The disasters to be considered vary with time and location and purpose of the underground space. Existing design codes evolve with disasters as more scientific knowledge and understanding is acquired. Some disasters can be prevented by valid geological prospecting, structure design, and construction management. These include structural disasters that lead to the damage of sidewalls, cracking, road arching, subsidence, dislocation, etc. Besides structural measures, many subsystems contribute to the antidisaster capacity of an underground space.

Not all disasters will be covered in this chapter; the focus will be on fire, inundation, and terrorist attack.

6.6.1 Fire

6.6.1.1 Fire Risk

Although the risk of a significant fire incident in road tunnels is low, it is still possible (Table 6.1). Underground fire incidents differ from fires that take place at the surface. Indeed, when a fire occurs in an underground

Table 6.1 List of road tunnel fires from 1974 to 1999

Year	Tunnel	Country	Tunnel length (km)	Fire duration	Damage		
					People	Vehicles	Structure
1974	Mont Blanc	France/Italy	11.6	15 min	1 injured	—	—
1976	Crossing BP	France	0.43	1 h	12 injured	1 truck	Serious
1978	Velsen	Netherlands	0.77	1 h 20 min	5 dead, 5 injured	4 trucks, 2 cars	Serious
1979	Nihonzaka	Japan	2.04	159 h	7 dead 2 injured	127 trucks, 46 cars	Serious
1980	Kajiwara	Japan	0.74	1 h 30 min	1 dead	2 trucks	Serious
1982	Caldecott	United States	1.03	2 h 40 min	7 dead 2 injured	3 trucks, 1 bus, 4 cars	Serious
1982	Lafontaine	Canada	1.39	—	1 dead	1 truck	Limited
1983	Pecorila Galleria	Italy	0.66	—	9 dead 22 injured	10 cars	Limited
1986	L'Arme	France	1.10	—	3 dead 5 injured	1 truck, 4 cars	Limited
1987	Gumefens	Switzerland	0.34	2 h	2 dead	2 trucks, 1 van	Slight
1990	Rødal	Norway	4.65	50 min	1 injured	—	Limited
1990	Mont Blanc	France/Italy	11. 6	—	2 injured	1 truck	Limited
1993	Serra Ripoli	Italy	0.44	2 h 30 min	4 dead 4 injured	5 trucks, 11 cars	Limited
1993	Hovden	Norway	1.29	1 h	5 injured	1 motorcycle, 2 cars	Limited
1994	Huguenot	South Africa	3.91	1 h	1 dead 28 injured	1 bus	Serious

(*Continued*)

Table 6.1 (Continued)

Year	Tunnel	Country	Tunnel length (km)	Fire duration	Damage		
					People	Vehicles	Structure
1995	Pfander	Austria	6.72	1 h	3 dead 4 injured	1 truck, 1 van, 1 car	Serious
1996	Isola delle Femmine	Italy	0.15	–	5 dead 20 injured	1 tanker, 1 bus, 18 cars	Serious
1999	Mont Blanc	France/Italy	11.6	2.2 days	39 dead	23 trucks, 10 cars, 1 motorcycle, 2 fire engines	Serious (closed for 3 years)
1999	Tauern	Austria	6.40	15 h	12 dead 49 injured	14 trucks, 26 cars	Serious (closed for 3 months)

Source: Data from ASHRAE Handbook used by Maevsky, I. Y. (2011). Chapter 4: Significant fire incidents in road tunnels—literature review. In *Design fires in road tunnels—A synthesis of highway practice*, NCHRP Synthesis 415 (pp. 21–26). Washington, DC.

space, imperfect combustion occurs due to the lack of oxygen, which produces larger amounts of smoke and poisonous gas. Since this occurs in a confined space, there is a limited capacity to remove heat and smoke, and temperatures rise quicker in underground spaces. Escape routes can also be narrow, long, and have limited visibility due to smoke. This increases the dependency on emergency lighting and evacuation signs. In addition, it is difficult for firefighters to judge the extent of a fire in an enclosed space. This means fire risk mitigation during the design phase is vital.

Although every tunnel or underground structure is unique, numerous design guidance standards and regulations have been developed around the world to ensure fire safety. In theory, the frequency of tunnel fires is related to variables such as tunnel length, traffic density, speed control, and road slope. Each influencing factor must therefore be taken into consideration when comparing different underground structures.

One of the biggest underground fire incidents was the 1999 Mont Blanc tunnel fire (Table 6.1). It became a catalyst for the improvement of fire safety in underground structures and especially tunnels (Hemming Fire, 2014). It caused the loss of 39 lives and major damage to the tunnel structure and equipment. It was then shut down for 3 years. While this fire was linked to an accident, some fires are intentionally started. One such example is the 2003 arson attack in Daegu, South Korea. In February 2003, a man set fire to a coach in a subway train using volatile materials. The death toll reached 192, and 148 were injured. It also led to property damage costing over 40 million US dollars (Hong, 2004). The tragedy revealed many flaws of the Daegu metro system, such as very poor visibility of the emergency lighting due to smoke obstruction and electricity power outage (Hong, 2004).

Fig. 6.2 shows the major causes of worldwide metro fires from 1903 to 2004 (Du, 2007). Two reasons that might not seem obvious are mechanical failure and electrical equipment. The former can be due to insufficient maintenance or poor build quality that may cause the derailment of a train and lead to a fire. But such cases generally cause smaller fires. However, electrical failures tend to cause significantly larger fires. A deadly example is the 1995 Azerbaijan metro fire disaster that killed around 300 people. It was caused by old wiring that sparked a fire (Reeves, 1995). Indeed, high-power cables are often used in underground structures and create risk as any short circuit, overload, or spark may cause a fire.

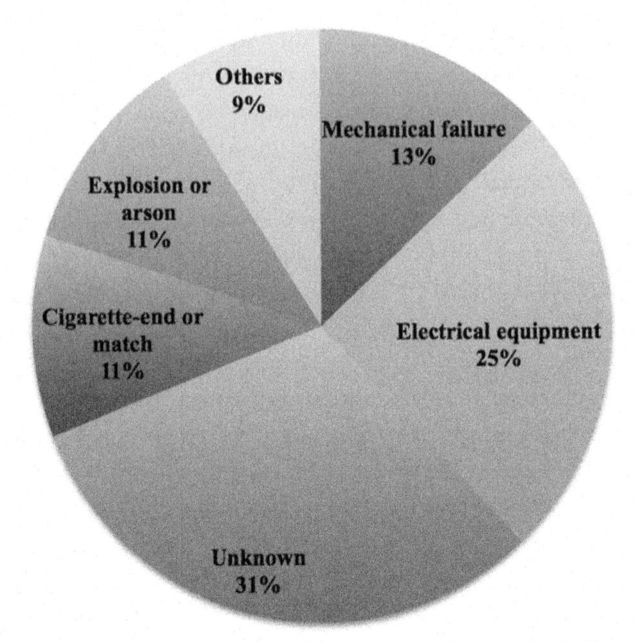

Figure 6.2 Causes for metro fire. *Data from Du, B. L. (2007). Statistic analysis of the foreign underground fire accidents cases.* Fire Science and Technology, *26(2), 214–217 (in Chinese).*

6.6.1.2 Fire Mitigation

Some factors that help ensure fire safety are (Han & Chen, 2006):

- Properly designed fireproof distance (minimum distance between adjacent buildings) and ground fire control facilities.
- Light wells and smoking shafts situated far enough from surface buildings to prevent fire from spreading.
- Space for children, elderly people, and disabled people not be placed underground if possible due to their reduced mobility.

It is also important to minimize fatalities in the case of accidents by having well-guided emergency exits, emergency equipment, and communication devices.

Fire compartments are separate spaces divided by firewalls that can stop a fire from spreading. The maximum area of such a space is defined by several factors including function, decorating materials, fire alarm, and extinguishing systems (Shao, 2007). Smoke compartments can also be designed using smoke screens, smoke beams, or smoke partitions. Smoke compartments are used to keep high-temperature flue gas in a specific zone so that it can be properly removed when activating the ventilation.

Another innovative measure to reduce causalities is the use of fire curtains (Fig. 6.3). At tunnel portals there may exist meteorological wind pressure, which can swiftly vary (Aigner Tunnel Technology, n.d.) and is a real issue. Fire curtains enable the reduction of the natural tunnel airflow and thus potentially enabling the tunnel to remain smoke-free and so users can safely exit the tunnel.

Egresses are often the escape path for users, the way smoke is vented, and the access path for firefighters. As fire may cause traffic chaos, entrance and exit design is critical. The following should be kept in mind (Liu, 2005):

1. Each fire compartment should have at least two exits facing different directions outside the building. For underground structures, which have several fire compartments, the fireproof door of an adjacent compartment can be considered as an exit but one egress should always be ensured.
2. Short escape routes should be designed for exits of underground shopping centers or entertainment places.

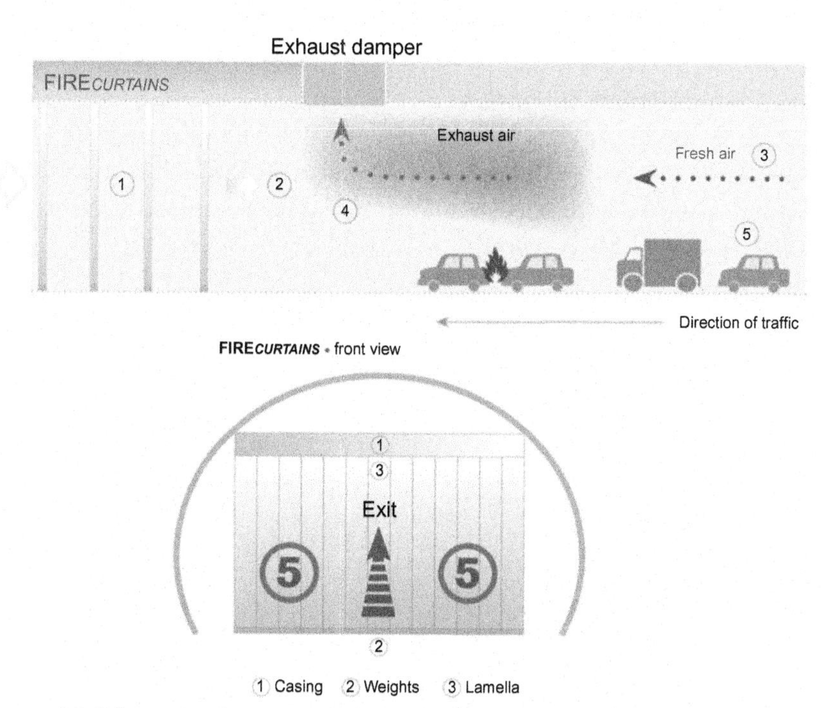

Figure 6.3 Schematic diagram of fire curtains. *Retrieved from Aigner Tunnel Technology. (n.d.).* Fire curtains. *Retrieved from <http://www.aignertunnel.com/index.cfm?seite = fire-curtains&sprache = EN>.*

Smoke-proof staircases and enclosed staircases should be used according to the industry's codes. For instance, fire-resistant separation is needed in stairs between the ground floor and the underground floor for people to escape and to prevent smoke from entering instead of going outside.

6.6.2 Inundation

6.6.2.1 Inundation Risk

Underground structures may also be part of a contingency plan in the case of flooding. A recent example is that of the stormwater management and road tunnel (SMART tunnel) in Kuala Lumpur, Malaysia. It can be used for water storage and transport. Nonetheless, water inundation is considered as another disaster that threatens underground spaces. It may cause problems not only during the construction stage but also during operation.

In the United States in 2012 Hurricane Sandy struck New York City. Although some preventive measures ahead had been taken, several subway tunnels and stations were flooded (Fig. 6.4). It was considered to be the most devastating disaster the more than a century old subway system had known (AFP, 2012).

6.6.2.2 Inundation Mitigation

Based on the experience gained from the 2007 flood in the New York area, three primary problems were identified in the event of an underground space being flooded (Kaufman, Qing, Levenson, & Hanson, 2012 based on a 2007 MTA report):

Figure 6.4 A surveillance camera image of water engulfing an underground station (AP, 2012).

1. lack of coordination among employees due to poor communication within the public transportation network;
2. insufficient facilities to prevent water from entering the system and to pump it out; and
3. lack of reliable information provided to passengers.

When a city's drainage system cannot prevent flooding, water may get into underground spaces such as metro systems. Although the severity of floods directly influences the damage caused, proper protection methods may help keep damage to a minimum. Inundation control is therefore of great importance.

Control methods to prevent flooding can include (Kaufman et al., 2012):

1. targeting of the most flood-prone locations;
2. installing valves to keep pumped-out water from re-entering the underground space;
3. improving sewers;
4. installing a Doppler radar in operations centers;
5. creating a public transportation system-wide emergency response center;
6. providing sufficient pump systems; and
7. developing operational strategies, including plans for predeploying portable pumps and personnel when storm conditions threaten to cause flooding.

Another way to avoid water from entering an underground structure is to raise the elevation of its accesses above the maximum flood level where possible (Bendelius, 1996). If this solution is not feasible, floodgates can be used (Fig. 6.5). They prevent water from flowing into the underground space. These steel gates are designed to resist hydraulic forces and are watertight. However, a potential drawback is that floodwater can be blocked at the portals. Sandbags are also an easy and relatively efficient method to divert water. Alternative technologies are also being developed, such as inflatable plugs (Sosa, Thompson, & Barbero, 2017).

A good drainage system limits flood damage as it keeps water from entering an underground structure. If water cannot be kept out of an underground structure, proper drainage is also needed to pump water out. This can be done in numerous ways depending on the design and number of pump stations located at low points.

One example of a drainage system can be seen in the Potomac River subway tunnel. The Potomac River is the fourth largest river along the

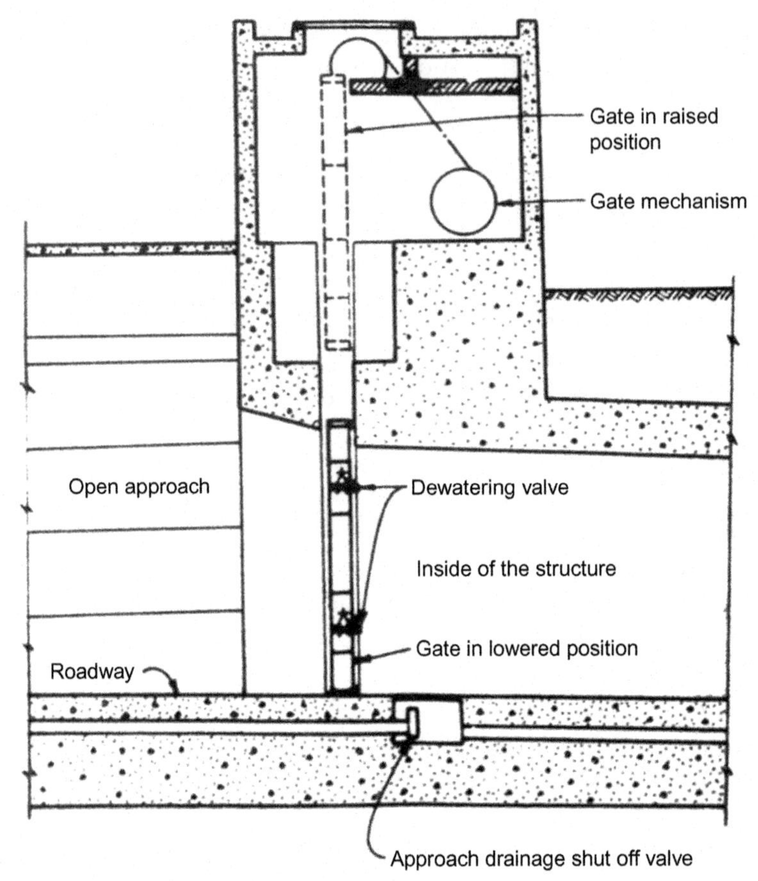

Figure 6.5 Typical floodgate installations at portal of tunnel (Bendelius, 1996).

Atlantic coast of the United States, spanning over 652 km from West Virginia to Maryland. The tunnel was bored in the 1970s through rock and is 1830 m long. Its drainage system consists of one low point as shown in Fig. 6.6. The water from the trackway is drained through embedded pipes to the low point station by gravity. There it is pumped to the surface by a pressure discharge line.

6.6.3 Terrorist Attack and Counterattack Strategies

6.6.3.1 Terrorist Attack Risk

Urban underground infrastructures are usually located beneath major cities. The high population concentration above increases the risk of underground structures becoming targets for terrorist attacks. Counterterrorism

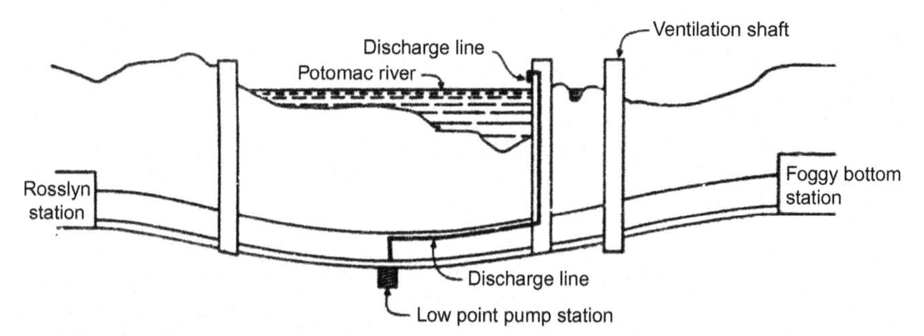

Figure 6.6 Profile of Potomac river crossing (Bendelius, 1996).

measures should therefore be considered. Terrorist attacks in underground structures are carried out in various forms, such as explosions and chemical attacks. Explosions are the most common tactic and of all the international terrorist attacks from 1968 to 1995 nearly half were bombings (National Research Council, 1995).

In 2010 in Moscow, suicide bombers carried out an attack in the metro system during the morning rush hour, killing 38 people and injuring even more (BBC, 2010). Biological and chemical attacks are also other forms of terrorist attacks. In Tokyo in 1995, for instance, members of a cult used the nerve agent sarin in the metro, leading to the deaths of 12 people and sickening thousands (Latson, 2015).

6.6.3.2 Terrorist Attack Mitigation

To cope with terror attacks, it is important to mitigate their consequences (O'Neill, Robinson, & Ingleton, 2012) to make them less attractive for assailants. This can be achieved by having well-established emergency procedures as well as disaster prevention devices, alarms, and rescue devices to minimize casualties.

Explosions must be considered in the design and dynamic analysis carried out. Understanding the physical phenomena and features of explosions can help to correctly define load cases and thus adopt efficient design features. A successful underground structure planning thus needs to be a collaborative work between structural engineers and explosive experts.

The European Commission has proposed a "Secure Metro" program that aims to increase resilience with vehicle design and thus reduce the consequences of such attacks on passengers, staff, and infrastructure (O'Neill et al., 2012). This plan also includes an in-depth analysis of perpetrated terrorist attacks to map particular trends (European Commission, n.d.).

As for structure planning, the following principles should be considered (Xi, Wang, & Zhang, 2005):

1. Explosions may lead to progressive collapse. The capacity of resisting progressive collapse can be improved by adding braces. Structures should be designed statically indeterminate to ensure overall stability when local failures occur.
2. In fire protection planning, engineers should consider local fire safety rules and standards.
3. Suicide attack often comes with a shock wave. Under the influence of the shock wave, splashes of fragments of glass and steel in the area of the explosion can cause serious casualties. To reduce losses caused by such secondary damage, special explosion-proof construction and decorating material is needed.

To counteract biological or chemical attacks, designers need to focus on ventilation, air condition, and smoke exhaust system integrating different features: blocking, fall blocking, filtration, discharging, and eliminating (Zhu, Zhu, & Li, 2005).

6.7 CASE STUDY: OPERATION STAGE OF SHANGHAI YANGTZE RIVER TUNNEL

This case study is an extract of an article published in 2012 by Bai and Liu (cf. References). After introducing the project, fire tests carried out for this tunnel were discussed before giving operation feedback 2 years after completion.

The Chinese coastal area is the most economically and industrially advanced region of Mainland China. The Yangtze River separates China's coastal area into the northern and southern part. The Shanghai Yangtze River Linkage project is located at the mouth of the Yangtze river and has provided another new traffic route to connect the northern and the southern coastal regions (Fig. 6.7).

China's state council decided to build the Shanghai Yangtze River Linkage in 2004. The total investment was about 2.06 billion euros, of which 1.26 billion euros was used for the linkage between Shanghai and Chongming Island.

The route for the linkage is located in the Yangtze River Delta. The maximum river depth is 20 m at high tide. The ground condition in the Shanghai area is very uniform. Shanghai's soil has a history of about only 6000 years and the rock bed is below 300–400 m. The site investigation

Figure 6.7 Map of China showing the Shanghai Yangtze River tunnel.

along the route of the linkage identified that the formation was alluvium, consisting mainly of soft silty clay, clayey silt, sandy silt, and fine sand.

The main geological layers (Fig. 6.8) are ①$_1$ alluvial clay; ②$_1$ sandy silt; ④$_1$ grey muddy clay; ⑤$_1$ grey muddy clay; ⑤$_2$ grey clayey silt with thin silty clay; ⑤$_3$ silty clay; ⑤$_4$ fine sand; ⑦$_1$ grey clay silt; ⑦$_2$ grey sandy silt. One can see that unfavorable geological conditions exist along the route of the tunnels, such as liquefied sand and silt, ground gas (methane), lenses and confined aquifer.

6.7.1 Fire Tests

Unlike rock tunnels under mountainous regions, any structural damage of soil tunnels under water may cause flood disasters over a wide area. As

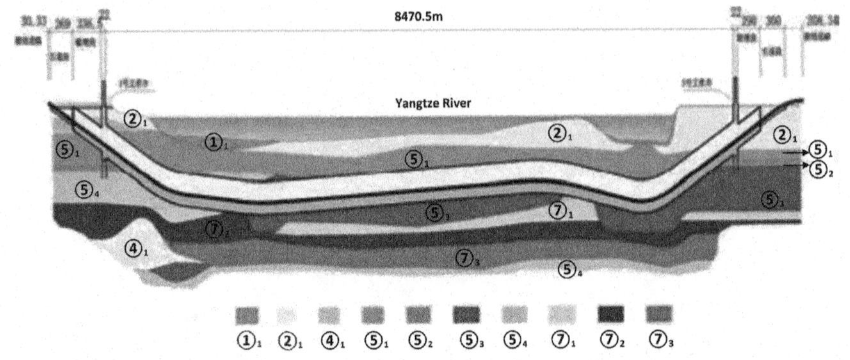

Figure 6.8 Geological section along the longitudinal profile of the tunnels.

Figure 6.9 Experiments at the fire site of the full-scale tunnel.

such, for this project hazard prevention (especially fire risk) was extensively considered. In order to check the design of ventilation, equipment, etc., a test tunnel was specifically built 100 m long, 10 m height, and 12.75 m wide. Four ventilators were installed: three for testing and one for standby. Each ventilator provided a power rate of 200 kW with the capacity of 125 m^3/s and was able to provide a maximum wind pressure of 900 Pa. Thus, the test tunnel had a wind velocity of over 4 m/s when three ventilators were in full operation. The test tunnel was located in the suburbs of Shanghai and, at the time, was the biggest test tunnel in China. During the test, a variety of pool fires, wood crib fires, and vehicle fires with different heat release rates were modeled (Fig. 6.9). The fire tests yielded the following results:

- In 20 MW tests, the highest temperature inside the test tunnel observed was 626°C.
- The foam-mist water system could extinguish 20 MW fire within 12 minutes.
- The fire alarm system responded within 30 seconds.
- Fire confirmation from the central control room was given within 60 seconds.

- The ventilator could start working steadily within 30 seconds.
- The optimum starting time for the auto extinguisher should be within 2 minutes.

6.7.2 Operation Feedback

In accordance with the principle of tunnel durability design a series of site measurements were conducted including tunnel settlement, cross-sectional deformation, and action of pressure on tunnel lining, rebar stress, and tunnel leakage. After 2 years of operation, the following feedback was given:

- The displacement between the rings is within 8 mm.
- The total upheave of the tunnel is within 30 mm.
- The total tunnel settlement is within 20 mm.
- The tunnel leakage has remained stable with a value of 0.0293–0.0290 L/(m^2 d), which is much less than 0.05 L/(m^2 d) of the design value.
- The maximum difference in tunnel load pressure between measurements and calculations is 31%.
- The maximum tunnel section deformation in terms of relative diameter deformation is only 0.45%.
- It was established from rebar stress measurements that the axial load force had become stable within 1 month of ring installation; these measurements are all larger than the design calculation.
- The traffic flow rate is still about 50% of the design capacity, which is 3055 passenger cars per tube per hour.
- No serious accidents occurred inside the tunnels.
- The ventilation result is satisfactory at today's traffic flow rate.

6.8 QUESTIONS

6.1. What are the main disasters in underground structures?

6.2. What are the consequences of a possible disaster and how can we prevent them effectively?

6.3. Describe the possible disasters and protections in underground engineering.

6.4. What is the function of cross-passages in road and metro tunnels?

6.5. Why do we need ventilation in a tunnel under construction or in operation?

REFERENCES

AFP. (2012). *Superstorm sandy: New York subway system flooded in 'worst ever disaster'.* Retrieved from <http://www.telegraph.co.uk/news/worldnews/northamerica/usa/9642268/Sandy-New-York-subway-system-flooded-in-worst-ever-disaster.html>.

Aigner Tunnel Technology. (n.d.). *Fire curtains.* Retrieved from <http://www.aignertunnel.com/index.cfm?seite = fire-curtains&sprache = EN>.

AP. (2012). *A still taken from a surveillance camera capturing footage of water engulfing an underground station* [Photograph]. Retrieved from <http://www.telegraph.co.uk/news/worldnews/northamerica/usa/9642268/Sandy-New-York-subway-system-flooded-in-worst-ever-disaster.html>.

Bai, Y., & Liu, Q. W. (2012). Shanghai Yangtze River Tunnel: Key issues in planning, design and construction. *Proceedings of 2012 Swiss tunnel congress*, 84–95.

BBC. (2010). *Moscow metro hit by deadly suicide bombings.* Retrieved from <http://news.bbc.co.uk/2/hi/8592190.stm>.

Bendelius, A. G. (1996). Water supply and drainage system. In J. O. Bickel, T. R. Kuesel, & E. H. King (Eds.), *Tunnel engineering handbook* (2nd ed., pp. 467–484). New York: Chapman & Hall.

Chinese Design Code for Underground Lightening [CECS45-92].

Du, B. L. (2007). Statistic analysis of the foreign underground fire accidents cases. *Fire Science and Technology, 26*(2), 214–217. (in Chinese).

European Commission. (n.d.). *SecureMetro report summary.* Retrieved from <http://cordis.europa.eu/result/rcn/140205_en.html>.

Han, F. Y., & Chen, H. (2006). Urban underground public building fire protection design. *Low Temperature Architecture Technology, 3*, 37–38. (in Chinese).

Hemming Fire. (2014). *Tunnel fires: Why they are vulnerable to disaster, the consequences and possible solutions.* Retrieved from <http://www.hemmingfire.com/news/fullstory.php/aid/2122/Tunnel_fires:_why_they_are_vulnerable_to_disaster,_the_consequences_and_possible_solutions_.html>.

Hong, W.H. (2004). The progress and controlling situation of Daegu subway fire disaster. In *Proceedings of the 6th Asia—Oceania Symposium on Fire Science and Technology*, Daegu, Korea, p. 28.

Kaufman, S., Qing, C., Levenson, N., & Hanson, M. (2012). *Transportation during and after hurricane Sandy.* Rudin Centre for Transportation, NYU Wagner Graduate School of Public Service.

Latson, J. (2015). *How a religious sect rooted in yoga became a terrorist group.* Retrieved from <http://time.com/3742241/tokyo-subway-attack-1995/>.

Li, J., Li, C., & Li, X. (2010). Test of air quality in subway tunnels. *International Society of Indoor Air Quality and Climate—ISIAQ.* Retrieved from <https://www.isiaq.org/docs/PDF%20Docs%20for%20Proceedings/2A.11.pdf>.

Liu, R. C. (2005). Fire protection design for underground shopping mall. *Fire Science and Technology, 24*, 34–38. (in Chinese).

Maevsky, I.Y. (2011). Chapter 4: Significant fire incidents in road tunnels—literature review. In *Design fires in road tunnels—A synthesis of highway practice, NCHRP Synthesis 415* (pp. 21–26). Washington, DC.

National Research Council. (1995). *Protecting buildings from bomb damage: Transfer of blast-effects mitigation technologies from military to civilian applications.* Washington, DC: The National Academies Press.

O'Neill, C., Robinson, A. M., & Ingleton, S. (2012). Mitigating the effects of firebomb and blast attacks on metro systems. *Procedia: Social and Behavioural Sciences, 48*, 2518–3527.

Park, S. (2016). *Back in the day—Daegu subway fire in 18 February 2003*. Retrieved from <http://www.knewsn.com/kn/bbs/board.php?bo_table = KNEWS&wr_id = 187>.

Reeves, P. (1995). *Old wiring caused worst metro disaster*. Retrieved from <http://www.independent.co.uk/news/old-wiring-caused-worst-metro-disaster-1580102.html>.

Shao, L. Q. (2007). Shallow talking about fire prevention design of underground building. *Fujian Construction Science & Technology, 1*, 66–67. (in Chinese).

Sosa, E. M., Thompson, G. J., & Barbero, E. (2017). Experimental investigation of initial deployment of inflatable structures for sealing of rail tunnels. *Tunnelling and Underground Space Technology, 69*, 37–51.

Xi, J. L., Wang, H. L., & Zhang, T. (2005). On the safety design and building measures for subway against terrorist attacks. *Modern Urban Research, 8*, 8–13. (in Chinese).

Zhang, Q. H., & Zhu, H. H. (2005). *Underground engineering*. Shanghai, China: Tongji University Press.

Zhu, P.G., Zhu, Y.X., & Li, X.F. (2005). Study on the security technique for the metro air environment under the nuclear biological or chemical terror. In *Proceedings of the 10th international conference on Indoor Air Quality and Climate*, Beijing, China, pp. 285–290.

INDEX

Note: Page numbers followed by "*f*" and "*t*" refer to figures and tables, respectively.

D

E

F

K

L

M

N

R

S

T

U

V

W

X

Z

Printed in the United States
By Bookmasters